# An Overwhelming Case For The Christian Worldview

**The Apologetic Series, Volume 1**

Justin Horn

Published by Justin Horn, 2024.

AN OVERWHELMING CASE FOR THE CHRISTIAN WORLDVIEW

**First edition. June 1, 2024.**

Copyright © 2024 Justin Horn.

ISBN: 979-8227021991

Written by Justin Horn.

# Table of Contents

# Table of Content

# Introduction

Science and Religion, are they really opposed to one another? Do scientific facts refute the idea that God exist? Or that the world was created in six literal days not very long ago? In this book I will attempt to show the reader that not only is Science and Religion compatible with one another but that in light of the evidence, the literal account of Genesis is the best explanation for the world around us. To this end I will be covering multiple scientific fields of study as well as philosophical and supernatural arguments.

To be clear I am not a "Scientists" per se, that is, I do not possess a degree in a science related field, however I will be referencing many works from those that do hold degrees. However, I urge the reader not to dismiss my claims just because I didn't go to college because that would be an...

**Appeal to Authority-** Believing something is true just because someone with credentials also thinks it's true. Example "Professor so and so says stars can form outta clumps of gas so it must be correct".

1Co 1:26 For ye see your calling, brethren, how that not many wise men after the flesh, not many mighty, not many noble, *are called:*

1Co 1:27 But God hath chosen the foolish things of the world to confound the wise; and God hath chosen the weak things of the world to confound the things which are mighty;

1Co 1:28 And base things of the world, and things which are despised, hath God chosen, *yea,* and things which are not, to bring to nought things that are:

1Co 1:29 That no flesh should glory in his presence.

I believe that for most if not all, the denial of God or belief in evolution is a heart issue, not a head issue. Meaning they don't want to believe in God because they like to indulge in their sin and they don't want anybody telling them what to do. If it was a head/intellectual issue then they would be open to the evidence.

Rom 1:18 For the wrath of God is revealed from heaven against all ungodliness and unrighteousness of men, who hold the truth in unrighteousness;

Rom 1:19 Because that which may be known of God is manifest in them; for God hath shewed *it* unto them.

Rom 1:20 For the invisible things of him from the creation of the world are clearly seen, being understood by the things that are made, *even* his eternal power and Godhead; so that they are without excuse:

Rom 1:21 Because that, when they knew God, they glorified *him* not as God, neither were thankful; but became vain in their imaginations, and their foolish heart was darkened.

Rom 1:22 Professing themselves to be wise, they became fools,

Rom 1:23 And changed the glory of the uncorruptible God into an image made like to corruptible man, and to birds, and fourfooted beasts, and creeping things.

Don't expect to convert an atheist simply by listing off scientific facts to them, it is by the preaching of the word of God (the gospel) that faith is obtained and one is saved, not by enticing words of man's wisdom. We cannot make someone believe in God, or surrender their life to Christ. That is between them and the Lord, however we can plant the seed and hope that it will grow; producing fruit.

Rom 10:17 So then faith *cometh* by hearing, and hearing by the word of God.

1Co 2:4 And my speech and my preaching *was* not with enticing words of man's wisdom, but in demonstration of the Spirit and of power:

1Co 2:5 That your faith should not stand in the wisdom of men, but in the power of God.

This book however is designed to stop the mouths of those who would dare claim that to trust in the bible is illogical, or that a belief in a young earth is unscientific. After reading this book you will be equipped to give an answer as to how you know Christianity is the way of truth.

1Pe 3:15 But sanctify the Lord God in your hearts: and *be* ready always to *give* an answer to every man that asketh you a reason of the hope that is in you with meekness and fear:

# Ch. 1. The Earth Bears Witness of the Earth Maker: Geological Evidence and Oceanography

## The Geologic Column

### 1. Poly-strate Fossils

The first piece of evidence that I'd like to present are poly (which means many or multiple) strata (as in strata, y'know those rock layers in the geologic column) fossils. I'm referring to tree trunks that can be seen running through multiple layers of sediment/rock. Now we have a dilemma for the evolutionist. If the layers of the geologic column was set down over millions of years then how do we explain these fossils. When a tree dies it falls over and rots away, there is no way a tree is just going to sit there and be covered by sediment for millions of years. Some of these trees are even found upside down. Obviously buried in Noah's flood. Creation 1 Evolution 0.

"Kent Hovind Creation Seminar 4, 35min"

### 2. Geologic Column's Missing Strata

The next piece of evidence that I would like to present is rather a lack of evidence for evolution. When a lot of us were little we were presented school textbooks with pictures of the so called "geologic column". This column presented a pile of layers; one on top of the other, with the oldest on the bottom and the youngest on top. All this spanning billions of years. Here is the problem. There is nowhere on earth where we find all of those rock layers put together like that except in the textbook. Most of the time they only find around 2 to 8 layers in a given area. Where did all the layers go?

The Evolution Handbook by Vance Ferrell pg 492-3

### 3. No Erosion between strata

If the rock layers where truly laid down slowly over millions of years we should expect to see some erosion, but we don't. This indicates they were laid down rapidly with very little time in between layers.

The Evolution Handbook by Vance Ferrell pg 632

## 4. "Cambrian Rock" Unconsolidated.

What do I mean by this point you ask? There exists sedimentary rocks even "Cambrian rocks", (FYI Cambrian rock is supposedly around 500 Million years old) that have been found to be in an unconsolidated state. This means that they haven't been pressed into solid rocks yet. Now if they have just been lying around for millions of years under millions of tons of rock there can be no doubt they would have been consolidated.

The Evolution Handbook by Vance Ferrell pg 662

## 5. Flood better explains geologic column.

Suppose the geologic column was a legitimate record of what we find in the rocks. It turns out a global flood would actually explain what we find in the fossil record better than the evolutionary scenario. Here is what happened during the global flood.

The first things that we would expect to be buried in a flood would be marine animals (which are found in "Cambrian sediment" and "Paleozoic gravel"). Then we would see some fish and the slowest land creatures covered in the "Silurian" dirt. Triassic soil would cover the slower of the dinosaurs, while the "Jurassic" and "Cretaceous" would cover the remaining Dino's. The quicker mammals would be covered in "Tertiary" sediments. We would not expect to find many birds or humans since they would be likely to find the highest altitudes and last the longest as well as cling to trees etc.

The Evolution Handbook by Vance Ferrell pg 686-687

## 6. The Lewis Over-thrusts.

What is an Over-thrust? An over-thrust is supposedly when older layers of rock for some reason move on top of younger rock. In another words when evolutionists find rock layers in the wrong order they just say one moved on top of the other. The Lewis over-thrust is a HUGE amount of rock 3 miles deep, 135 miles long, apparently slide sideways for many miles over undisturbed shale. Now shale will crumble if put under sideways pressure. So how did all that land move over that shale without demolishing it? Heart Mountain is another example of an "Over-thrusts". C'mon guys, we all know those rocks aren't millions of years apart. When you have to move a mountain to save a theory, it's time for a new theory.

The Evolution Handbook by Vance Ferrell pg 506-9, 504

## 7. Geosynclines (Bent Rock layers).

Sometimes people have discovered bent rock strata commonly called Geosynclines. These folded and/or buckled rock strata range in size from considerably to miles long (these ones are folded mountains). This is no problem for Creationists as Noah's flood could easily explain this phenomenon. Obviously these folds would have to have been made while the sediment was still soft and pliable. "Millions of years old rock" would simple crack and break off under such pressure.

The Evolution Handbook by Vance Ferrell pg 499-501

## 8. Footprints.

Tourists have been flocking to Glen Rose of the Paluxy River Valley for decades. Located in Texas, dinosaur track ways have been excavated by depression era-locals (looking for some extra cash), as well as real Paleontologists. What makes this destination so controversial is of the supposed human tracks located in the same strata as Dino tracks. This has been claimed for a long time by some creation scientists and some local residents as well. Because of the uncertainty of the evidence and also because much of what was once there has been removed and cannot be examined, answers in genesis does not believe that coexisting Dino and human prints are completely acceptable. Dr. John Morris does not believe Creationists should use the Paluxy evidence against evolution any longer, unless new facts are brought to light.

https://answersingenesis.org/dinosaurs/footprints/paluxy-river-tracks-in-texas-spotlight/

## 9. Lack of bioturbation between layers.

If Evolution were true and the earth was billions or even just millions of years old, then we should expect to see bioturbaiton between the layers of the geologic column, but we don't. What is bioturbation? Bioturbation is the disturbance of the sediments by plants and animals. Shucks, we don't even find soil between the layers.

https://www.creationsciencetoday.com/
17-Lack-Soil_Layers-Bioturbation.html

# Refuting "Millions of Years" Formation

## 10. Earth's magnetic field

Earth's magnetic field has been deteriorating rapidly. Precise measurements have been going on all over the world way back since 1829 A.D. 20,000 years ago the Earth's magnetic field would have been comparable to the sun's, this equals no life 20,000 years ago; hence no evolution. The magnetic field will supposedly be gone by 10,000 A.D. Hence no life because of cosmic rays.

https://www.allaboutcreation.org/age-of-the-earth-c.htm

## 11. Rapid forming opals.

Opals are a precious gem, but instead of taking a crystalline shape, opals instead are amorphous. Evolutionist teach that these gems take hundreds of thousands of years to form. However according to Australian researches opals most likely can form in a matter of either weeks or months. They concluded this by inference by studying known growth rates for bacteria and compared that to the fossil bacteria which grew alongside the opals in order to estimate the time taken to for the opals to develop. I apologize if that sounded confusing and if I didn't articulate that well, here is the link that hopefully explains it better.

While we are on the topic of precious gems, it turns out that gemstones prove the inspiration of scripture, and thus God. Tiny fragments of various gemstones have been examined under the microscope using Cross-polarized light. It turns out that the gemstones mentioned as being used in the making of New Jerusalem are Anisotropic, meaning they are beautiful and colorful under Cross polarized (pure) light; whereas diamonds, rubies, and garnets are Isotropic jewels, meaning they are nothing to look at under that type of light.

https://www.icr.org/article/opals-can-form-weeks/

Youtube Video: New Jerusalem by A&O Productions

## 12. Rapid forming oil.

The commonly held belief is that Oil is said to take a vast amount of time to be made. It is even claimed to take million of years. Is this true?, if so it would seem to discredit the bible. Thankfully this is not the case. It turns out that oil taking long ages to form can be dis-proven in multiple ways.

1. Oil can be created from sewage in 24-48hrs, although this is not naturally occurring, hence the next point.
2. Petroleum was created in the lab with conditions simulating nature.
3. There is a natural refinery beneath the sea off the gulf of California that is producing oil now. This oil has been C-14 dated to be around 4200 to 4900 years old.

https://creation.com/how-fast-can-oil-form

## 13. Coal.

Coal is a fossil fuel, which is an organic deposit that produces energy when burned. Many believe that coal takes long periods of time to form. It turns out however that laboratory experimentation has shown that coal can form in hours or days under the right conditions. It also turns out that we know of natural phenomenon that produce coal rapidly as well. During the aftermath of the Mt. St. Helen's eruption a log was discovered with wood on one end and the other side contained coal. So it turns out that creationist have nothing to fear about coal, as coal does not contradict the young earth model of the bible.

https://www.icr.org/article/origin-coal

## 14. Halo's.

According to Evolutionary theory the granites took millions of years to cool and form from hot magma, is there any evidence this is not the case. You guessed it, there sure is. It turns out the granites contain polonium halo's (magnificent microsperes of coloration). However if the granite took so long to form we would not find these things in there. Consider this allagory, what happens when we drop an Alka-Seltzer in water? It fizzes for a short time and then is quickly gone. The only way to preserve this effect of fizzing would be to instantly or near instantly freeze the glass. So to the granites must have formed instantly in order to preserve these marvelous evidences of creation.

https://www.halos.com[1]

## 15. The Lost Squadron.

When skeptics took Kent Hovind to a Colorado freezer to show him Ice core rings, from cores they got in green land. They told Kent that they had seen a core with 135000 rings (which supposedly represented summer/winter). Actually they weren't annual rings, that's an assumption. They represented hot, cold, hot, cold. As proof look up the lost squadron. In ww2 (1942) some planes ran out of fuel and had to stop in Greenland. They forgot about them, and then 48 years later someone decided to go find them. They were under 263 feet of ice. The deepest hole they drilled was 10,000 ft, divide that by 5.5 a year (taken from how long the lost squadron was under the ice) and we get around 1800 years.

Evolution Handbook pg 678, " Kent Hovind Creation Seminar 1, 1r 28min"

## 16. Surtsey.

---

1. https://www.halos.com/

Surtsey is just one example of how quickly land can form. Surtsey is an Island which was created in a matter of days in the north Atlantic from an underwater volcanic eruption in 1963.

Surtsey features:

- Lagoons

- sandy beaches

- spectacular cliffs

- gravel banks & much more

It turns out that this Island (and others as well) prove that earth can form quickly under the right circumstances, therefore there is no reason to assume or attribute deep time to geological rock formations.

https://creation.com/age-of-the-earth

https://creation.com/surtsey-the-young-island-that-looks-old

## 17. Mt. Saint Helen's and sediment build up.

Evolutionists would have you believe that the grand canyon as well as other rock layers took millions of years to form, after all time is the hero for the evolutionary model to work. There are multiple examples from the 1980's Mt. St. Helen's eruptions that prove otherwise. The rock layers were formed by catastrophism, not long gradual ages. I will give three examples derived from the following link.

1. Put shortly the Mt. St. Helen's eruption is what caused over a million logs to be pulled into spirit lake, and within half a decade the bark from those logs (mixed with volcanic sediment) had formed a peat layer over three ft thick.

2. There were various effects of the aforementioned eruption(s) including landslides, volcanic ash from the ground and air, mud-flow, and waves from spirit lake. All of which drastically changed the environment. These different forces when added together resulted in

up to 600 ft thick complex sediment layers. The Mt. St. Helen eruption also created volcanic ash beds ranging in size from a fraction of an inch to three feet, all in several minutes or less. Multiple layers were laid down as a result of this catastrophe. In June 12, 1980, during the course of only a few hours, a sedimentary layer was formed with a size of 25 ft thick. It also contained various sediment layers.

3. It was the Mt. St. Helen's eruptions that formed the "Little Grand Canyon of the Toutle River". This miniature grand canyon (which factually formed very rapidly compared to the evolution modal), is a one-fortieth scale modal of the real grand canyon.

https://answersingenesis.org/geology/four-lessons-mount-st-helens-eruption/

## 18. Radiometric Dated Mt. St . Helen's.

From 1980 to 1986 Mt. St Helen's has erupted multiple times, this produced Dacite lava which formed the dome shaped cap at the eruption site. When the Dacite was collected in June of 1992 and taken to a lab, the ten year sample was 340,000 to 2.8 Million years old. Showing once again we cannot put our faith in men's dating methods.

https://answersingenesis.org/geology/radiometric-dating/
radio-dating-in-rubble/

## 19. C14 in Coal

Plants take in Carbon 14, we eat the plants or we eat the animals that eat the plants, so everything has this carbon 14 in their system. Then when we die, we slowly lose this C-14. This is because C-14 is unstable (radioactive) and converts to nitrogen. It is important that the reader notes the following—-there should be no Carbon-14 in samples older than 100 thousand years. Coal is *supposed* to be _300 Million years old_. Yet we find measurable amounts of C-14 in it.

CrEvo Rant #101 with Wazooloo

## 20. C 14 in Diamonds.

Why are diamonds so valuable? Some will tell you it's because they take millions or billions of years to form. Is this true? No, the simple fact that diamonds contain C-14 demonstrate that diamonds cannot be that old. Since diamonds are so hard it is highly unlikely that the evolutionist argument for contamination is valid.

"Ultimate Proof for Creation/God's Existence" by Jason Lisle

## 21.Stalactite/Stalagmite growth

Have you ever visited a cave? If so, you may have heard the tour guide state that these Stalactites/Stalagmites have formed slowly over several millions of years. Now is that necessarily true? If so the bible is in error and we cannot trust it. Thankfully this is not the case. It turns out Stalactites and Stalagmites can form rather rapidly given the right conditions. Want some examples try researching Lincoln memorial stalactites, Stalactites on Philadelphia bridges, bat encased in stalagmite.

Kent Hovind Creation Seminar 1 "The Age of the Earth", https://creation.com/rapid-stalactites

## 22. Bell and Pot found in Coal

Self-Explanatory, but I will go ahead and explain anyway. Ask any evolutionist and he will tell you coal takes many millions of years to form. We Creationist would say most or a lot of the coal was laid down by Noah's flood roughly 4200 years ago. According to evolution, humans didn't arrive on the scene till a couple million years ago or so. So how do we find human artifacts inside these lumps of coal.

"Kent Hovind Creation Seminar 2 The Garden of Eden"

## 23. Megabreccias

What on earth is a megabreccia? A megabreccia is an enormous boulder that has been moved to it's current location by massive floodwaters. The sheer weight of these rocks is astounding! One example of such megabreccas are found in Peru, located in Eocene strata a very long distance from their original position. So what was their weight? These huge megabreccias weighed up to 5,000 metric tonnes (the weight in pounds is 11 million). Each boulder has a diameter of 32.8 to 49.2 feet. These seem like very good evidence of a global flood.

The Evolution Handbook by Vance Ferrell pg. 501

## 24. Fossilized Hat.

If you ever researched how long long does it take to make a fossil, the search results may show it takes at least 10,000 years for a fossil to form and could take millions of years. Again how do they know this? Where they there when the fossil was forming, not if it took millions of years like they say. Below I list a link where you can see a picture of a miner's hat that turned to stone in what I presume less than 100 years. There is even a fossil pickle, acorns, and teddy bear that you can see just by watching Seminar 6 the Hovind Theory by Kent Hovind. In the video I saw it was around about the 1 hr 6 min mark. I would also like to add you can find a picture of a petrified cowboy boot with the foot still inside on Seminar 1, 1hr 33min mark.

https://answersingenesis.org/geology/catastrophism/fossil-hat/

## 25. Dating Methods give maximum ages not precise ages.

When using dating methods in order to determine the age of a specimen, it is NOT like looking at a clock. The time (age) is. Instead dating methods are like looking at an hourglass. It gives _up to _ages not exact ages.

If we enter a room with an hour glass that has 2/3rds of the sand at the bottom (supposedly 40 minutes have past). We have to make the assumption though (see next point) that nobody has came in the room and turned it over before we got there though.

## 26. Dating Method ASSUMPTIONS.

Time and time again evolutionists tout dating methods as proof of evolution (or rather old ages). They do this in order to discredit the bible's modal of a 6 to 10 thousand year old earth. But can we really rely on these dating methods to give us reliable information? The short answer is no, and the answer is because scientists must make assumptions when dating any given sample, some of theses assumptions include:

- The initial conditions (no daughter elements present at the start).

- The rate of decay has not changed.

- The atmosphere was the same in the past as it is today.

- Each sample was in a closed system (It would be Ideal that the sample was protected from outer space radiation, water, chemicals, etc for however long the sample has existed).

The Evolution Handbook by Vance Ferrell pg 168-171

# Water and Oceanography

## 27. Colorado River.

Evolutionist will tell us that the Colorado river formed the grand canyon over millions of years. Is this a fact or is this an assumption? It is an assumption, as my sources indicate the Colorado river did NOT form the grand canyon over millions of years. Any geologist worth his salt would tell you that slow moving winding rivers do not cut through hard rock, they can only cut through soft rock. The Evolution Handbook seems to imply that the grand canyon was very rapidly cut through when it's layers had just been laid; and the Creation seminar tells us that the grand canyon is a giant washed out spillway, and that the enormous barbed canyons on the east side verify that the grand canyon is a breech in a giant dam.

The Evolution Handbook by Vance Ferrell pg 663, Kent Hovind Creation Seminar 4, 13min

## 28. Niagara Falls Erosion.

In 1678 French Explorer Henepin mapped the falls. From the point it was discovered until 1842 it was calculated the Niagara falls eroded the underlying cliff at 7ft per year. Although more recent calculations say it's about half that. That would land the Niagara Falls at an age of five to ten thousand years old. That however is not factoring in Noah's flood which would have really done a lot of the eroding of the initial conditions.

The Evolution Handbook by Vance Ferrell pg 145

## 29. The Law of Mass Action

The common belief among evolutionists is that complex chemicals in the oceans came together to form enzymes, then those enzymes got together to form proteins, polysaccharides, lipids and nucleic acids. Aside from the impossibility that random chemicals could produce amino acids out of mere chance (see point 76).

If enzymes or proteins could theoretically form themselves out of some primitive organic soup from the ocean all the excess water would have to be then immediately removed or else those enzymes and/or proteins would break down before they could do anything. The majority of what we find in sea water is just that, sea water. There isn't enough concentration of chemicals in the ocean to form amino acids.

The main point is chemical compounds are meant to be inside of us, when they leave the body they are rapidly destroyed.

The Evolution Handbook by Vance Ferrell pg 216-17

## 30. Geologic layers are formed by water not long ages, it's science.

Time for a science experiment, grab a glass mason jar, or some other type of jar with a lid. Now fill it with water, gravel, sand, rocks, dirt, and mud. Now shake it vigorously for let's say 1 minute. Now see how long it takes for the contents to sort themselves out in layers. Hint: not millions of years. In moving water it turns out that we can get multiple layers to settle at the same time.

Kent Hovind Creation Seminar 4, 24min & 30min

## 31. The amount of Salt in the Ocean.

The ocean is getting saltier every year, that means that the input and output of sodium entering and leaving the ocean hasn't yet reached equilibrium. Evolutionist claim the ocean is 3 Billion years old. The Maximum age of the ocean based on the input and output of sodium is in the 40 to 60 million year range based on the assumption there was no salt in the beginning and then going by the current rate. This doesn't take into account the sodium gain from the volcanic dust and massive erosion during the great flood.

https://www.icr.org/article/oceans-salt-clock-shows-young-world

## 32. Amount of sediment on sea floor.

Magnanese Nodules. Very few of them are found in the deeper layers of sediment of the sea floor. Although the rates of sea sediment accumulation are very slow as evolutionist would expect, this does not explain why we do not find very many Magnanese Nodules in sediment deeper that 20 inches. This would be expected in the Creationists model, because the deeper layers were laid down by the global flood very rapidly; not giving the Nodules time to form. Some may argue that the Manganese Nodules dissolve with depth but this doesn't seem to be the case.

https://www.icr.org/article/8503

# More Problems with Deep Time

## 33. Meteorite Dust

Every day about 20 million meteors collide with our atmosphere, these meteors increase earths weight by 25 tonnes daily. If we compare the present rate of meteor gain to the amount previously deposited in the strata we get an age of thousands of years. Indicating a young earth.

The Evolution Handbook by Vance Ferrell pg 136

## 34. Meteor Craters.

No meteor craters are found in the lower rocks. All meteor craters are on or near our surface. The absence of craters in the rock strata indicate no large meteor impacts older than a few thousands years ago.

The Evolution Handbook by Vance Ferrell pg 137

## 35. Meteorite Rocks.

Likewise, if meteors have been bombarding the earth for millions of years, then it stands to reason that we would find billions of meteorites in the rock strata, but we never find any in the lower, supposedly "older" layers. Asimov's theory of crustal mixing tries to explain this away but cannot account for the absence of nickel in the earth strata.

The Evolution Handbook by Vance Ferrell pg 137

## 36. Oil Pressure.

When well diggers penetrate the rock and strike oil, a geyser of oil springs toward the sky. This gusher is caused because the trapped oil was under an immense amount of pressure. However according to studies of the permeability of the surrounding rock, that pressure should have dissipated after a few thousands of years. And yet evolution would have you believe that oil formed slowly over vast amount of time.

The Evolution Handbook by Vance Ferrell pg 142

## 37. Oil Seepage.

In addition to oil pressure, oil seepage also provide a dilemma for deep time. While the oil seepage rate from the sea floor is not given in my source, it gives a maximum age of twenty thousands years; after that all the oil would be gone. It turns out however that geologist have found 630 BILLION barrels of oil that can be recovered from these off shore wells.

The Evolution Handbook by Vance Ferrell pg 142

## 38. Topsoil.

Worldwide, topsoil averages about 8 inches. Even allowing for erosion, it has been calculated that the earth gains an inch every three hundred to one thousand years. This is consistent with the biblical account of a young earth.

The Evolution Handbook by Vance Ferrell pg 145

## 39. Helium in Zircon Crystals.

Zircon crystals were retrieved from a very deep and very hot hole in New Mexico. It is known that heat increases chemical activity. Amazingly however these Zircon Crystals still contained helium. Helium is a gas, which would have diffused out if these crystals were much older than a few thousands years.

The Evolution Handbook by Vance Ferrell pg 144

# Possible Biblical Sites and Other Interesting Finds

## 40. City of Sodom Found?

I will let the reader judge for themselves whether this evidence is true or not. I would not want to mislead anybody. The only reason I say this is because the discovery was made by a man named Ron Wyatt (who is not a certified archaeologists) who I mention elsewhere may not be credible. On the other hand, who's to say that God cannot use an uncertified person to excavate and uncover biblical artifacts? God is sovereign and often uses the weak things of this world to confound the mighty, scripture also says that God uses the base things and the despised things and the things that are not to bring to nothing the things that are. So that no flesh should glory in his presence. 1 Corinthians 1:27-29. So I'm not ruling out that Ron Wyatt's discoveries are legitimate. Anyway what discoveries were made that prove or rather lend credence to the existence of the cities of Sodom and Gomorrah?

On the following link you can find photographic evidence of...

· Sulfur Balls, supposedly 95 or 96% pure sulfur.

· Multi-layered Ash (evidence of extreme heat) comprised of Calcium carbonate & sulfate; which are by products of the reactions of sulfur and limestone burning.

· Formations that are apparently man-made (Spinx's, windows, ziggarates, arched doorways, and walls.

· Sulfur Crystals.

"https://www.arkdiscovery.com/sodom_&_gomorrah.htm"

## 41. Mt. Sinai Found?

Well, sort of. According to the got questions link provided there are multiple places that are proposed to be the real Mt. Sinai. In case your new to Christianity, Sinai or Mt. Horeb was the place were God gave the ten commandments to Moses. One candidate for Mt. Sinai is located on the south Central part of the Sinai Peninsula, this is where Constantine sent his mother (Helena) after a dream supposedly informed him this location is the true Mt. Sinai.

Another, much stronger candidate for the real Sinai (in my opinion) is the mountain located at the northwestern Saudi Arabia at Jebel el-Lawz. Although I haven't done a lot of research on the real location of Mt. Sinai, the 2nd link which covers the latter site in some detail contains multiple pictures including a <u>burnt mountain top, a carving of a calf in rock</u> (possibly made when the Israelites went a-whoring after the golden calf), and a giant split stone. These pictures are some of the strongest evidence that I'm aware of that this is the true Mt. Sinai.

Exo 19:18 And mount Sinai was altogether on a smoke, because the LORD descended upon it in fire: and the smoke thereof ascended as the smoke of a furnace, and the whole mount quaked greatly.

Exo 32:4 And he received *them* at their hand, and fashioned it with a graving tool, after he had made it a molten calf: and they said, These *be* thy gods, O Israel, which brought thee up out of the land of Egypt.

Exo 17:6 Behold, I will stand before thee there upon the rock in Horeb; and thou shalt smite the rock, and there shall come water out of it, that the people may drink. And Moses did so in the sight of the elders of Israel.

https://www.gotquestions.org/mount-Sinai.html

https://evidence-for-the-bible.com/archeological-evidence-for-the-bible/archeological-evidence-for-mount-sinai-jebel-el-lawz/

## 42. Split rock Found?

I realize that I just covered this in my last point, however I though that I'd add an extra link that goes into great detail concerning his magnificent landmark. There are actually a water erosion channel between the split in the rock that descends from the top of the hill. Also a relevant thing that I'd thought that I would bring up is that Moses wasn't allowed into the promised land because he smote the rock twice. This destroyed important symbolism since that rock represented Christ.

Exo 17:6 Behold, I will stand before thee there upon the rock in Horeb; and thou shalt smite the rock, and there shall come water out of it, that the people may drink. And Moses did so in the sight of the elders of Israel.

Num 20:11 And Moses lifted up his hand, and with his rod he smote the rock twice: and the water came out abundantly, and the congregation drank, and their beasts *also*.

Psa 78:16 He brought streams also out of the rock, and caused waters to run down like rivers.

Psa 78:20 Behold, he smote the rock, that the waters gushed out, and the streams overflowed; can he give bread also? can he provide flesh for his people?

1Co 10:4 And did all drink the same spiritual drink: for they drank of that spiritual Rock that followed them: and that Rock was Christ.

Heb 10:14 For by one offering he hath perfected for ever them that are sanctified.

https://www.truth-absolute.com/found—split-rock.html[2]

## 43. Air bubbles in Ancient Amber sample contain 50% more Oxygen than Current Atmosphere.

This is evidence that the evolutionary assumption of "the present is the key to the past" aka uniformatarianism is not valid.

"Kent Hovind Creation Seminar 2: The Garden of Eden"

## 44. Flat Plateaux (Elevated Paleo plains).

Some of these Plateaux are dated many millions of years old. Gondwana Surface of southern Africa has been dated to the Cretaceous era, and some paleo plains of central and west Australia have been assigned to the Triassic era. How these plateforms have survived without being destroyed several times over is a mystery according to Davisian theory, but poses no problem for young earth creationists.

https://creation.com/age-of-the-earth

https://www.icr.org/article/did-landscapes-evolve

---

2. https://www.truth-absolute.com/found--split-rock.html

# Ch. 2 Bones Deficiencies in the Evolutionary Theory: Evidence from Paleontology

## Frauds and Hoaxes (Paleontology)

**45. Lucy.**

As I am sure many people know, Lucy is one of the most popular supposed missing links in history. It may or may not come to the surprise of the reader that some evolutionists really don't care if to support lying if it furthers the cause of evolutionary indoctrination. Such is the case with the St. Louis Zoo located in Missouri, United States. Even though Bruce Carr (Zoo Director of Education knows the exhibit is flawed (because the feet were inaccurate), he refused to change the exhibit because he believes the exhibit's overall impression is correct.

The bones of Lucy do not match up with human bones at all, the only reason evolutionists have for claiming this specimen as a human missing link was the Laetoli tracks. The laetoli tracks appear human because they are human tracks, the reason they are claimed to be a missing link or attributed to Lucy is because they are found in strata that would buck against the assigned ages of evolutionary strata.

https://answersingenesis.org/human-evolution/lucy/ape-woman-statue-misleads-public-anatomy-professor/

https://www.icr.org/article/more-evidence-lucy-was-ape

**46. Piltdown Man Fraud**

Here is another supposed missing link in human evolution that was found. It took almost forty years for the scientific community to recognize this an one giant hoax! How do we know that Piltdown man was a fraud?

· A good sized portion of Piltdown man was fabricated from plaster.

· Sure there were parts of a human Skull but also...

· It contained the Jawbone and teeth (which were filed down) of a chimp!

· It was stained to appear ancient, additionally there were imported mammalian bones and tools nearby.

"Kent Hovind Creation Seminar 2, 59:20", The Evolution Handbook by Vance Ferrell 533-535

## 47. Nebraska Man Fraud

Here is yet another supposed evidence of a human missing link, and what is this amazing evidence that proves evolution to be true without a shadow of a doubt. A SINGLE tooth, a molar to be precise. This tooth of "Nebraska man" was a key piece of evidence during the scopes trial and was said to be more human than ape. It turns out it was neither human nor ape, not even close. It was a PIGS TOOTH!

Kent Hovind Creation Seminar 2, 58:44 mark, and The Evolution Handbook by Vance Ferrell pg. 536

## 48. Archaeorapter Fraud.

It turns out that in China one can make quite a bit of money selling bird fossils, despite them being real or not. It is quite easy to create counterfeit rock and adhesive, but extremely hard to spot. National Geographic actually confirms Archaeorapter is a hoax. It is a chimera of two different creatures. A tail from the Dromeaosaur and a bird's body.

https://answersingenesis.org/dinosaurs/feathers/archaeoraptor-hoax-update-national-geographic-recants/

## 49. Java man hoax.

The bones of this supposed missing link was found in Indonesia in the very late 1800's. It is comprised on a skull cap, a leg bone and some teeth. Never mind that the leg bone was 45 feet away from the skullcap. The leg bone was human but the skull cap was a little different from the usually human type. It was later discovered that *Pithecanthropus erectus* or 'up-right ape man' was found in the same strata as regular people so he was just another type of the human kind, not a missing link. Also note that *Pithecanthropus erectus* was renamed *homo erectus 'up-right human'.*

https://creation.com/not-another-ape-man

## 50. Neanderthal man fraud.

Among the list of supposedly found missing links of human evolution we come to the Neanderthal. Is the Neanderthal some sort of half ape, half human transitional form. Nope. The differences between Neanderthal bones and humans bones can easily be accounted for by diseases such as arthritis, rickets, and congenital syphilis (*this is thought to be the case by some but see point #63*).

Additionally, there was a Neanderthal skeleton found in the Philippines, with that amount of moisture in the region, it couldn't be even a century old. There was also found a Neaderthal skeleton with unrusted chain-mail armor. Oh, and Neanderthals had bigger craniums/brains than us. That would be the opposite of upward and on-wards evolution. The elongated cranial vault and heavy brows are from advanced age, the bible tells us that people lived for hundreds of years prior to and just after the flood.

The Evolution Handbook by Vance Ferrell pg 526-28

## 51. Orce Man.

This supposed ancient ancestor of man turned out to be just a skull fragment of a four month old donkey.

The Evolution Handbook by Vance Ferrell pg 602

## 52. Ape and Monkey Skulls.

There is something an evolutionists needs to consider. What is that? It's the fact that Immature ape skulls look strikingly similar to human skulls, and what's more; if there were giant monkeys in the past their skulls would also look quite human like and could easily be mistaken for some kinda missing link. We know for a fact that animals were much bigger in the past (see point 54), therefore it is not unreasonable to assume there were larger monkeys.

The Evolution Handbook by Vance Ferrell pg 552

**53."Cavemen" Misconception.**

One misconception that we may get from public school is that ancient cultures lived in caves and then over time as men became more "evolved" we slowly advanced from these cave cultures to more complex and sophisticated housing structures. That is one way to look at it, but here is the problem; there are societies today that live in caves. There are cultures in various parts of the globe that have differing levels of advancement. This does not prove that one culture evolved into another, furthermore this has nothing to do with the notion that one animal changes into another. When people move from one location to another it may take them a while to build houses, in the mean time they may choose to live in caves or other "primitive" structures. It's also a possibility that cultures can become less advanced over time.

The Evolution Handbook by Vance Ferrell pg 523-526

# Troublesome Discoveries for Evolutionist

**54. Giant Critter Fossils**

For this section I would like to give just a small amount of samples of giant animals that lived in the past.

- **Giant Dragonfly 50 inch wingspan.**

- **Cockroaches were 18 inches**

- **2ft Grasshoppers**

- **Centipede 8.5feet**

- **Donkey 9ft tall**

- **60ft Cattails**

- **Elk with Antlers of 12 ft**

- **1500lb Guini Pig fossil**

- **10ft Kangaroos**

- **and Oysters 11.5 ft wide!**

The implications of this in regard to evolutionary theory is huge. This is because Evolution teaches animals are getting bigger, and stronger, and faster, and smarter. This is not what we find in the fossils, we find that we are going in the opposite direction that evolution predicts. This answers the notion of how man could've lived around dinosaurs and not went extict. One possibility is that mankind was bigger also in the past. Another possibility as we will find out later, is that mankind was a lot smarter in the past (see point 179).

Gen 6:4 There were giants in the earth in those days; and also after that, when the sons of God came in unto the daughters of men, and they bare *children* to them, the same *became* mighty men which *were* of old, men of renown.

Kent Hovind Creation Seminar 2 1hr 17min

## 55. Mammoth.

It's leg carbon dated 15380(RCY), it's skin and flesh was 21300 (radio carbon years).

Kent Hovind Creation Seminar 1, 1hr 22min

## 56. "Cambrian Explosion".

For this small section I will ask the reader one question. Which creation modal better fits the Supposed "Cambrian Explosion"? Darwin's tree of life has simple cell organisms turning into ever-more complex organism as a trunk of a tree then over millions of years as we go up the metaphorical tree it branches off into different kinds of animals, then over more millions of years off of those branches we have twigs of different species and sub species. According to Genesis though all the different kinds of animals were there at the beginning. The Cambrian Strata that is supposedly 500+ million years old has all or nearly all the major phyla of animals arriving on the earth from the beginning or near the beginning.

The Case for a Creator by Lee Strobel pg42-45

## 57. Hell Creek Formation, Montana.

T-rex buried with clams, sharks, turtles, and Amphibians.

https://www.icr.org/article/a-trex-swimming-with-sharks

## 58. Real Dragon Skull Dracorex Hogwartsia

(Named after abominable Harry Potter Franchise)

I've decided to include this example to show the reader that dragons did indeed exists. The word dinosaur wasn't coined until the 1800's after all. Now I'm well aware that the fairy-tale version of dragons probably doesn't exist. That is because it is a chimera (an incorporation of characteristics from multiple dinosaurs). The body of a T-Rex, tail of a stegosaurus, Skull similar to a triceratops, wings of a Pterodactyl, but I digress. The Komodo dragon is like a modern day dinosaur, if not in fact is one. I highly recommend the following link if you want to learn more about real historical dragon accounts.

https://www.genesispark.com/exhibits/evidence/historical/dragons/

## 59. Smithsonian Cover-up.

It turns out that the Smithsonian is responsible for hiding giant human skeletons, though some may deny this, this author believes this to be true. We already know that some evolutionists do not care to misrepresent evidence aka lie in order to protect their beloved theory. A giant human skeleton would definitely refute the evolutionary dogma that we started off small and are getting bigger, and stronger, etc. There is evidence in my source that some humans grew to over 12 feet. Giant human skeletons would also confirm multiple bible verses about giants as well.

"Kent Hovind Creation Seminar 2, 55min"

## 60. Trilobites.

Well what's this about trilobites? Do they also debunk evolution? They do in fact. A trilobite was found inside a human shoe print, located in what evolutionists consider Cambrian rock (shale). Discovered by William Meister and verified by Utah's geological survey.

" Kent Hovind Creation Seminar 4, 31min"

## 61. Down-washing & Reworking

Down-washing and Reworking are two more ways that the geologic column is proved to be incorrect. Evolutionists frequently find bones lower and higher in the "geological record" than they are supposed to be by their evolutionary dogma. Either the evolutionists will ignore the evidence or they will say that the bones/fossils somehow moved through solid rock in order to get to the "wrong strata". In the case of down-washing an animal was "washed down" to a rock layer which was supposedly older than the time that it was alive. In the case of reworking the animal fossils' surrounding host rock was eroded and then the fossil was incorporated into newer/younger rock. These are nothing more than rescuing devices deployed by the evolutionists so they don't have to give up their worldview.

The Evolution Handbook by Vance Ferrell pg 491

## 62. Shells with Amino Acids.

How long do you think amino acids can exists for? A thousand years, a million? How about 135 to 180 MILLION YEARS! Such would be the case if evolution were true. Shells from the Jurassic strata have been discovered containing (inside of them) amino acids locked into the proteins. The links some people would go to hold on to a crumbling theory.

The Evolution Handbook by Vance Ferrell pg 484.

## 63. Neanderthal Skulls

After studying the mouth palate of ancient skulls, Dr. Jack Cuozzo (works in Paleoanthropology, which is the studying of ancient man) discovered that kids of old time developed (matured) much slower than children today. This is his first retro-diction (a prediction about the past). This would be expected if humans lived hundreds of years in the past like the bible says.

Evolutionary Paleoanthropologists knew that Neanderthals bones show signs of slow growth, longevity, as well as slow maturation, but try and cover it up. Some attribute differences in Neanderthal bones to disease or nutritional deficiencies but Jack believes these differences are on account of old age.

**Arthritis** does not make neanderthal bones from people who have died recently.

For debunking the **rickets** argument Jack shows examples of Neanderthal skulls that do not possess soft calcification and/or box like heads (which are two signs of rickets). Additionally, we do not see high arched palates in Neanderthal youth palates in the examples Jack gives. Another thing Jack highlights about neanderthal skulls is that they have slow closing sutures instead of premature closing sutures. This is the exact opposite of what we would expect from a person who has rickets. Jack also found no decay in Neanderthal teeth as well (another sign neanderthals did not have rickets). Jack provides evidence of baby teeth lasting an unusually long time by modern standards as well (indicating slow maturation of youth).

Neanderthal skulls do not show the signs of **congenital syphilis** (notched incisors, little enamel left on mulberry molars, barrel shaped bottom incisors).

From the video provided we see small a face on neanderthal youth skull indicating slow development/growth in the past. Evolutionists would have you believe we evolved from apes that have quick growth rates then over millions of years growth rates slowed down as they evolved into man but in reality it's the opposite. We are maturing way faster than our ancestors did (oh, and we didn't come from apes). Dr. Couzoo witnessed a paleontologists cut the chin off of specimen La Quina V to make it more apelike, and then use dental compound to make it even more apelike. Evolutionists can't find evidence for their theory so they have to make it.

Dr. Cuozoo also provides X-ray evidence of a skull of a 2+year old that has the angulated palate of a fetus (another sign of slow maturation). It is 4 standard deviations below normal for a two year old, but that was a normal child.

He also discovered that somebody was washing off red ocre off of skulls because this tradition couldn't have been hundreds of thousands of years old, but Jack found some on the inside of the skulls. Red ocre is used by morticians to make the dead appear more alive.

Youtube: The Truth about Ancient Man Dr. Jack Cuozoo

## 64. Clams.

That were petrified were found on Mt. Everest, additionally they had to have been buried alive because when a clam dies it opens up, but these clam were in the closed position.

" Kent Hovind Creation Seminar 2" 1hr 21m

## 65. C-14 in Dino bones.

The mere fact that Carbon 14 is found in the bones of dinosaurs prove they are not millions of years old, since there should be no detectable amounts of C14 in a sample older than 100,000 years. Don't however expect an evolutionist to give up their worldview so easily under such evidence however, as they will most likely argue that the sample was contaminated. Belief in the Creator is a heart issue, not a head issue. Meaning there is ample evidence for God. Evolutionists don't _want_ to believe because that means they would have to obey him and give up their sin.

Carbon 14-Dating & The Age of Dinosaur Fossils | CREATION with David Rives

## 66. Human remains in old layers.

If the geologic column really did form over Billions of years we shouldn't expect to find human remains in the older layers. According to evolutionary theory dinosaurs didn't live with man, and yet we find human skeletons in Cretaceous rock (dated 70 to 135 million years ago)! That was the last rock layer that dinosaurs are supposedly found in. What's more, in the example they give in my source, there are 2 skeletons (the Moab skeletons) from Moab, Utah; the bones were still articulated, undisturbed, this is an indication that there was no earth movement.

There was another instance where ten, yes ten skeletons (male, female, and kids) were found in hard sandstone. The rock was gradually removed over the course of many years from mining operations. The skeletons were found under 50 feet of hard sandstone, the skeletons had been dyed blue and green from exposure to minerals. According to the geological survey the area they were found in was 100 million years old. According to Evolutionists they were either...

#1 Indian burials (why would Indians dig under 50ft of solid rock though, plus there was zero evidence of this supposed digging).

#2 The bones tumbled down some crack in the rocks. (Wouldn't the man operating the bulldozer have seen such a large crack).

#3 There was mining for copper done by Indians, then the cave collapsed trapping them. (No evidence of the tunnel, plus why would children and women be mining for copper?). This is not a very good theory as the bones were trapped in solid rock, not a collapsed tunnel.

Evolution Handbook pg 558-560

Youtube Video: OOParts special Genesis week Episode with Wazooloo, aka Ian Juby

## 67. Petrified cowboy boot with human foot inside.

" Kent Hovind Creation Seminar 4, 39min

## 68. Dinosaur Graveyards.

Here is a quote from Edwin Colbert, *Men and dinosaurs* (1968) p.141

"As the layer [cut out of a New Mexico hillside] was exposed, it revealed a most remarkable dinosaurian graveyard in which there were literally scores of skeletons, one on top of another and interlaced with one another. It would appear than some catastrophe had overtaken these dinosaurs, so that hey all died together and were buried together."

These piles of dinosaurs bones lean more heavily to a water catastrophe than to the dinosaurs dying over millions of years.

The Evolution Handbook by Vance Ferrell pg 666-677

## 69. Stasis in Fossil Record.

According to evolutionary theory animals change into other kinds of animals over vast ages of time. If this were true however why do we have so many examples of animals in the fossil record that have not changed at all in supposedly tens if not hundreds of millions of years? If it takes so long for an animal to evolve, then does it even happen at all?

## 70. Sharks.

**Sharks are a problem for evolutionary theory in two ways.**

**1.** First, they are a strong evidence of stasis. These animals have remained virtually unchanged for supposedly _over 400 Million years_. Now if evolution is an undirected process caused by mutations, what can cause sharks to just stop evolving? It's not like an animal can just choose what to change into or refrain from changing into. There are in fact no evidence of macro-evolution in the fossil record, all we see is stasis and extinction; neither of those helps evolution.

**2.** Another way sharks disprove evolution is from my source listed at the bottom of this point. It says that they have found sharks squished down to a quarter of an inch with the tail still upright. This would seem to me to be a good indication of a catastrophic flood.

The Evolution Handbook by Vance Ferrell pg 469.

## 71. Fossil Ichthyosaur Fish giving Birth

" Kent Hovind Creation Seminar 4, 39min"

# Where Are All the Bones?

# (+1 Logical Fallacy)

## 72. Bone Inventory.

So why does bone inventory support the creation modal over the evolution one? Because there is so little evidence that supports the notion that mankind has been on earth for millions of years. If we have been on earth this long then why do we have only around 1400 specimens that are used as examples of ancient man? What's more is that these are just scraps of bone and teeth. The whole amount can be placed easily inside just one coffin. Where are all the bones? This would be what we would expect to find if we have only been here a few thousand years.

The Evolution Handbook by Vance Ferrell pg 549

## 73. No Transitional Fossils says the experts.

It's one thing when a creationists claims that there are no transitional fossils. It's quite another when Evolution Scientists say that they cannot find and/or do not have any transitional forms. Without transitional forms evolutionists have no good case that macro-evolution has happened in the past. Let's look at three quotes of experts that claims there is no transitional forms found in the fossil record at present.

1. **Dr. Colin Patterson** (of the British Museum of Natural History). When questioned as to why he didn't include a picture in his 1978 book titled Evolution, he said... "I fully agree with your comments on the lack of direct illustration of evolutionary transitions in my book. If I knew of any, fossil or living, I would certainly have included them...

"[Steven] Gould [of Harvard] and the American Museum people are hard to contradict when they say there are no transitional fossils. As a Paleontologists myself, i am much occupied with the philosophical problems of identifying ancestral forms in the fossil record. You say that I should at least 'show a photo of the fossil from which each type of organism was derived.' I will lay it on the line——there is not one such fossil for which one could make a watertight argument."

Dr. Colin Patterson, letter that was dated April 10$^{th}$ 1979 to Luther Sunderland. Quoted from L.D. Sunderland's Darwin's Enigma p. 89.

1. **Dr. Niles Eldredge** (of the American Museum of Natural History, New York) "No one has found any such in-between creatures. This was long chalked up to 'gaps' in the fossil records, gaps that proponents of gradualism [gradual evolutionary change from species to species] confidently expected to fill in someday when rock strata of the proper antiquity were eventually located. But all the fossil evidence to date has failed to turn up any such missing links."A Niles Eldredge quote from "Alternate Theory of Evolution Considered," in Los Angeles Times, Nov 19$^{th}$ 1978.

2. **Dr. David Raup** (the field Museum of Natural history, Chicago) " We are now about 120 years after Darwin, and Knowledge of the fossil record has been greatly expanded. We now have a quarter of a million fossil species but the situation hasn't changed much. The record of evolution is still surprisingly jerky and, ironically, we have e en fewer examples of evolutionary transition than we had in Darwin's time!" David Raup "Conflicts between Darwin and Paleontology", Field Museum of Natural History Bulletin Jan. 1979.

These three men know what they are talking about, they have spent a lifetime examining fossils before saying those things. Between the three of them they have been in charge of over half of the major fossil collection on earth.

The Evolution Handbook by Vance Ferrell pg 459-62

## 74. Circular reasoning

Another problem for evolution is the fact that evolutionists date the strata by the index fossils they contain, and then go and date the fossils by which strata they are located in. It is reasoning in a circle. This is what's known as a logical fallacy.

The Evolution Handbook by Vance Ferrell pg 433

# Ch. 3 The Study of Life directs to the Life-Giver: Biological/Chemical Evidence

## The Impossibility of Chemical Evolution

**75. The Miller Experiment.**

So what about the miller experiment proves creation? Well, mainly I list this as proof of creation because it demonstrates the problems of getting life from non-life. So what are the problems with the Miller experiment?

1.  First you have to _assume_ the right atmospherical conditions of the early earth. The consensus today (of when my source was written) is that Miller did not use the right atmospherical conditions. Miller had a simulation of a hydrogen rich mixture of ammonia, methane, and water vapor. There was no evidence in favor of this supposed early earth atmosphere but a lot of evidence against it. According to newer theories of scientists of earth's early atmosphere, it contained only a small amount of hydrogen, but was mostly comprised of nitrogen, water vapor, and carbon dioxide. I'm no scientists but I personally believe the atmosphere had more oxygen than it does today (see points #43 and #180.

2.  What happens when you use a simulation of the "correct" atmosphere believed by modern scientists. Not amino acids but organic molecules. Well that sounds promising. Could these organic molecules be the precursor of life. Nope. Why? Because it's Formaldehyde Cyanide! The stuff is so toxic that the mere thought of it being used to create life is ridiculous at face value.

The Case for a Creator by Lee Strobel 37-38

**76. Amino Acids**

Evolutionary scientists speak of probabilities as if they were actually possibilities, if just given enough time. "Just give enough time and evolution can do the miraculous"! This is simply not true. Just because someone can calculate the odds of someone being able to jump across the widest part of the grand canyon it doesn't make it a real world possibility. It's the same with being able to throw a rock to the moon, the probability certainly can be wrote out on paper as a number with 50 or so zero's following it; but this doesn't mean that it is actually possible. Here are some big numbers to assist the reader of what I'm getting at.

$10^{18}$-is ten billion years in seconds.

$10^{28}$-diameter of the universe in inches.

$10^{80}$-subatomic particles in the universe.

Now do you wanna guess the chances that random processes could produce enough left amino acids just to form one protein molecule? If you guessed $10^{210}$ you'd be correct.

The Evolution Handbook by Vance Ferrell pg 266-69

## 77. Facts about Protein

What's the odds of making 124 specifically ordered proteins of 400 amino acids by sheer chance? It's $1X10^{64489}$! That is a gigantic number. It would take a book with over 60 pages to just write all the zero's down (providing 1000 zeros per page). So what are the odds of these 124 specific proteins are made up of only left handed amino acids (the only kind used to make a living organism) occurring by random processes? You would need 10 billions years, and have $10^{31}$ earths, with all their oceans consisting only of amino acids linking up every second of that time, plus they kept linking in sets of 124 proteins. The chances of that happening (get ready for it), $1X10^{78436}$. This is just to get to 124 proteins. We would need success in these odds times a MILLION in order to make evolution even possible.

The Evolution Handbook by Vance Ferrell pg 270-71

## 78. Enzymes

What problem do Enzymes pose for evolutionary theory. Well according to Fred Hoyle we need 2000 different enzymes (that are very complex mind you) to get a living organism. He also stated that even in twenty billion years (which exceeds the supposed age of the universe) we wouldn't get even one of these enzymes by the act of random shuffling.

Another thing that makes evolution a far fetched fairy-tale for adults is the fact that you need all (or nearly all) of the Enzymes of a system there before the system works. We also need to have pre-existing enzymes in order to make enzymes.

The Evolution Handbook by Vance Ferrell pg 271-72

## 79. The oxygen dilemma!

Amino acids can't form with oxygen, yet would be destroyed by cosmic rays otherwise (without it).

The Evolution Handbook by Vance Ferrell pg 223-225

## 80. The law of mass action

Would have hindered protein formation in the sea, the amino acids would have dispersed.

The Evolution Handbook by Vance Ferrell pg 216-217

# DNA & Microbiology

## 81. Hemoglobin Molecule

What are the odds of getting just 1 hemoglobin molecule by chance? By the way hemoglobin contains 574 specifically coded acids! According to S.W Fox and J.F. Foster this is what we need.

· Large random amounts of all 20 Basic protein molecules. In order to get this amount of protein molecules we gotta have a volume of $10^{512}$ x our Universe' s volume.

· They would have to be the left-handed ones (which occurs only half the time in lab experimentation. That is the only way random chance would produce a single hemoglobin molecule.

And then we still have to answer how we got all the other thousands of kinds of protein molecules found in every living cell!

The Evolution Handbook by Vance Ferrell pg 273

## 82. Laminin.

...is the protein network foundation for nearly all the organs and cells that makeup our bodies. Laminin is shaped a lot like a cross. You know, I bet there is a bible verse relevant to this.

Col 1:17 And he is before all things, and by him all things consist.

https://www.youtube.com/shorts/CpF86o9dc2A

## 83. No Junk DNA.

Up until relatively recently 98% of our DNA was claimed to be considered "Junk DNA" because it had no known function. If this were true it would definitely be an argument against intelligent design. However, thankfully this is not the case. Our DNA code is amazing complicated and it has been discovered that , those parts of the DNA do serve a purpose. To put it simply that "Junk" DNA serves as instructions on how to use other parts of the DNA code. The source I give goes into a great deal of more detail than I am going to mention here feel free to check it out, it's worth the look.

Genesis Week Season 2 Episode 22

## 84. DNA Translators.

And here is yet another dilemma for those who insist evolution is the truth. Not only do we need DNA to arise by chance spontaneously but simultaneously there had to be the machinery in place to translate that DNA onto the living tissue or else it would result in immediate extermination.

Evolution Handbook pg 248-249

## 85. Chicken and the egg scenario

DNA, proteins, enzymes they need each-other to exists. In order to have DNA, you gotta have proteins (which are enzymes), but in order to have proteins you need DNA. They had to come together all at once. That is quite an insurmountable hurdle for evolutionary theory.

The Evolution Handbook by Vance Ferrell pg 272

## 86. Mitochondrial DNA proves literal eve.

It turns out that even evolutionists will agree with creationists that a genetic bottleneck has occurred in the recent pasts, through testing of genetic samples (though evolutionists interpret the evidence based on evolutionary timescales, so they get much older ages of let's say 100,000 years). However when we study real-world examples of rates of genetic change (the studies of the bodies of executed Russian Tsars, for example), we get rates that are much much faster than initially assumed. It turns out that the dates from observed rates of change were as much as 20x faster than the rates given for the evolutionists assumptions to work!

"Genesis Week Season 3 episode 12. 22 min

## 87. The 99% myth.

There is a common belief pushed on society by evolutionists that humans and chimps share 99% of our DNA. This was taught way back in 1975 before it was possible to even compare the base pairs of the DNA of the human and chimp.

This Lie has continued even until recent times; *Science* (one of the top science Journals in the world) in a 2012 report was still claiming the 99% similarity myth.

Even giving evolutionist there 99% similarity we would still have a **thirty million** letters difference in our DNA! But what is the real percentage? How much DNA do we actually share with chimps. According to Drs Jeffrey Tomkins and Jerry Bergman reviewed and published studies the figures are at the most 87% similarity and it is possible that we are not even greater than 81% similar. This figure was the result of comparing all the DNA, not just some of it. According to Dr. Tompkins research we are only 70% similar.

https://creation.com/do-humans-and-chimps-share-99-of-dna

## 88. Reproduction 1

There is no good explanation for the origin of sex under the evolutionists model. Most reproductive systems of the millions of species of animals will not be compatible with each other. Yet evolutionists teach we have come from a common ancestor. The supposed first life forms on the planet (bacteria), reproduce asexually, they can generate huge population in hours. There is no good reason why natural selection should produce a much more intricate and risky system. Bacteria don't have to search around and find a mate.

Evolutionists believe many if not all land animals came from a fish ancestor. Fish lay eggs, then the male does his thing (to put it delicately for young audiences), the young are left to take care of themselves. Fished supposedly evolved into amphibians (different reproduction system), then into lizards (again, different reproduction system). Most lizards copulate in the same manner we humans do. Unlike fish, lizard eggs must be fertilized prior to the development of the shell. Under evolutionary theory animals of different species must develop these very complex systems (the male and the female) independently of each other, at almost exactly the same time and place on planet earth, the systems have to be compatible and the animals have to know how to use their equipment. Any number of things could go wrong (in a random universe) which would mean the species extinction. This takes a lot more faith to believe than the author possesses.

Another conundrum of evolution is the "male bone" as we will call it (just in case a young reader sees this). Most male mammals and all male great apes have one, so how did mankind's supposed ape-like ancestor simultaneously lose his male bone and at the same time develop a hydrolic system in order to mate? If this didn't happen (according to evolution) he would have no descendants and thusly we wouldn't be here.

Evolutionists may argue that if the bible were true then Adam and Eve's offspring would have to commit incest to reproduce. While that is true, incest was not dangerous until our DNA has deteriorated later on down the line. Evolution has the same issue, it also would requires incest multiple times for multiple generations. Even small problems with the reproductive system of creatures of the same species can result in no reproduction. So are we really believe that animals gradually and simultaneously made changes to their reproduction systems over millions of years.

Gen 1:27 So God created man in his *own* image, in the image of God created he him; male and female created he them.

CrEvo Rant #13 by Wazooloo (Mature Audiences only)

## 89. The Central Dogma.

... is a scientific principle and spelled out simply is this; what happens to an organism's body or in it's body cannot produce new species, only a change in the DNA can do that!

The Evolution Handbook by Vance Ferrell pg 279-280

## 90. Antibiotic resistant bacteria

...are the result of a loss of information not a gain, deformed ribosomes.

"Kent Hovind Creation Seminar 4, 52min"

## 91. Complexity of cell.

Although I do not agree with the old earth assumption that my source says (that 100 million years was the time elapsed between the earth "cooling" and the first micro-fossils), because obviously the bible says the earth formed out of water not lava, plus y'know polonium halo's, n'all. Anyway I found an interesting probability on the page from my source that I thought that I should include in my book. Know we already covered probabilities of amino acids forming by chance as well as proteins. Now we have another dilemma for evolutionary theory. If by some miracle we managed to get those amino acids and proteins by chance, then how many do you think we need (proteins that is) to get a minimally complex cell? 20? 50? 100? Try **300** to **500!**

"The Case for a Creator by Lee Strobel pg 229"

## 92. Bacterial flagellum.

The Bacterial flagellum is what we call a teleological argument, or argument from design. This bacterial hair is surprisingly complex. It is comprised of rods, rings, a rotor, a hook, filament, proteins, etc. This hair is attached to a motor. It also turns out that we could fit eight million of these motors on the cross section of a human's hair. It rotates at an incredible 100,000 rpm. Even more interesting is that smaller organisms have to deal with greater viscoscities than larger organisms. This means the liquid around them feels more sticky. Traveling through water for them is equivalent to a human swimming through peanut butter, Oh and I also need to mention that is would be like the human going at 60mph through that food stuff as well.

"Kent Hovind Kent Seminar 4, 1hr 49min"

## 93. Gene depletion

There is a principle in biology known as gene depletion, gene depletion says that when you selectively breed an organism, it will weaken it, not strengthen it. It is a loss of fitness through adaptation. The original stock of a given kind was resilient, and tough, but as it branched into different variations it became more feeble. That is gene depletion. (see point 121). What further harms an organism and increases the genetic load are mutations that build up over generations (see next point for a longer explanation). If evolution was a fact we would actually have less potential in our genes than our "animal ancestors" did!

The Evolution Handbook by Vance Ferrell pg 391-92

## 94. Genetic Entropy.

Evolution teaches that we get beneficial mutations in our DNA, this over time supposedly creates new organisms. For the sake of this example we are going to pretend this is true. We are also going to pretend that lethal mutations are fictional for the sake of this example. Now on to the example. Over the coarse of your life you build up mutations in your DNA, let's say about 100 of them in all. These are neutral, almost neutral, or bad mutations, but you also get one beneficial mutations as well. You then pass all these mutations on to your offspring; the 1 beneficial mutation and the 99 bad, near neutral or neutral ones. This cycle repeats itself for each new generation. This wouldn't be super bad unless humankind has been around for a very very long time as evolution supposes. This is because as each new generation acquires more and more genetic errors it makes the DNA more and more unreadable. Much like that of a copy of a copy of a copy of a copy and so on... of a tape recording.

"The boat that don't float: CrEvo Rant #78 with wazoolo"

## 95. Irreducible Complexity.

What is Irreducible Complexity? Basically, it tells us that living organisms must have a certain number parts right there at the beginning in order for it to survive/function/work. For the cell, it has to have a place to store the DNA, parts to let it take in food and excrete waste, parts to make it move around, lipids to form the cell membrane, etc. All these and more had to be there from the start or else the organism would not have survived. I like the example that Micheal Behe provides, the mousetrap. In order for the popular model that we usually see as a mousetrap to work it needs:

1. The wooden platform
2. The Metal Hammer that kills the rodent
3. The spring

4. The catch that releases at a tiny amount of pressure
5. The bar that attaches to the catch and holds back the hammer when trap is charged

Suggested Reading "The Case for a Creator by Lee Strobel" pg 195-201

# How the Animal Kingdom Rebels Against Evolution

## 96. Orderliness of Nature

For this point we are going to be covering the uniformity of nature as an argument against evolution. I left a recommended video at the end of this point that touches briefly on this topic, but for this section I'd thought that I would chime in with my own thoughts on the subject. If the universe was brought about by chance and there is no one overseeing the cosmos, then how come we see such order in the universe. By looking at the world around us we see such complexity and structure. There are many, many examples of this some of which are listed in this book. Elsewhere I list termite mounds and Globular clusters as evidence for the creator; though they also be examples of the order we see in nature. Other examples include bird migration, Bee colonies, Bird mating dances, symbiotic relationships, animal behaviors, the various components of a cell interacting one with another and performing specific functions, if I wanted to take the time I could probably give many more examples but that should suffice to make the point. If the big bang theory was true then why don't we have random crazy stuff just happen all around us. Could hitting our thumb with a hammer actually feel pleasurable? Could lava spewing from a volcano suddenly turn into cotton candy? Why don't we have a day where the laws of gravity gets suspended for a few hours? Why couldn't these things happen if the laws of nature where brought about by random processes?

"Ultimate Proof for Creation/God's Existence Youtube video by Jason Lisle"

## 97. Mankind is not living longer than ever before.

According to evolutionary theory (which is the theory that violates the 2nd law of thermodynamics), we are getting stronger, and faster, and smarter, and better. You may have been told by some that mankind is living longer now than ever before in earth's history, and that a few hundred years ago people lived much shorter lifespans. The first book of the bible (Genesis) list people as living over 900 years old. Some may argue that the years of those days were more like months or something 900 divided by twelve is 75 years. There are two problems with that theory;

1. 75 years old is about the average life expectancy today, so that proves we are not living longer.

2. More importantly, if each year was a month in Genesis five you would have Enoch and Mahalaleel only being <u>five years old </u>when they had their first sons.

(see Psa 90:ʼ10 & Genesis 5).

For additional information consider "Hovind Seminar 1, 1hr mark"

## 98. Natural Selection & Survival of the Fittest.

Some evolutionist call variation in offspring of organisms, **Natural selection**, and this by itself, they claim is the reason for all the various life forms on earth. One such example of an evolutionist that claims this is the case is Sir Julian Huxley, he states "So far as we know...natural selection...is the only effective agency of evolution." The name 'natural selection' should be called 'natural randomness', because evolution is supposed to happen by random chance variation. SELECTION indicates someone to do the selecting, which the theory of evolution denies. Randomness would not select only beneficial traits and move only in positive directions. The nearest thing to natural selection that we have is gene reshuffling, but this does not go across the species barrier to create a new species. (I personally think my source should have said kind here because species isn't always interchangeable with kind in every instance, it is kind that never changes).

Neo-Darwinist believe that natural selection alone could not produce cross species changes, therefore they rely on mutations (see point #139) to do the job.

**Survival of the Fittest-** is the culling process of weaker organisms that were the products of accidents or mutations. This culling out process actually moves the species closer to the original pattern, it does not produce cross species change.

Evolutionists say that evolution is a one way street that once changed, cannot go back to the former type. For example Chickens supposedly evolved from dinosaurs, but a chicken will never turn back into a dinosaur. To support this ridiculous theory they (evolutionists) point to Darwin's finches or pigeons. Micro-evolutionary changes that go both ways, not just one direction. Changes back and forth of beak shape and feather color cannot be extrapolated to say reptiles turn into birds. After he wrote his famous book origin of the Species, Darwin said to a buddy: "In fact the belief in Natural Selection must at present be grounded entirely on general considerations [faith and theorizing]...When we descend to details, we can prove that no one species has changed...nor can we prove that the supposed changes are beneficial, which is the groundwork for the theory. Nor can we explain why some species have changed and others have not." Letter to Jeremy Bentham.

It's true we have animal hybrids of closely related species of animals such as a horse and a zebra, a turkey and a chicken, etc; however the result of which causes the offspring to be either sterile or (the females descendants of such cross breeding) to produce young that do not live long.

The Evolution Handbook by Vance Ferrell pgs 284-285, 287, 294,296, 299

## 99. Panda's and Fruit Bats.

So how do Panda's and Fruit bats defy evolution? Well It turns out that panda's and fruit bats have sharp pointy teeth (the same kind of teeth that are said to indicate that an animal is carnivorous). But it turns out that both of these animals are herbivorous. You see the bible says that in the beginning there was no death, and that the Lord gave the animals the green herb to eat. Evolution teaches that many of the pointy teeth dinosaurs were carnivorous. The point that I'm making is that just because an animal has pointy teeth it does not meat that they have to eat meat, the Lord could cause them to eat plants like he did in the beginning. (see also point 162)

Kent Hovind Creation Seminar 2, 1 hr 43 min

## 100. Reproduction 2.

According to evolution we are living in a by chance and purposeless universe, how is it then that when we look at biology we see such amazing complexity in life forms, consider all the steps that must occur in order to produce a living child.

1. Every month when an egg is released, the follicle that contained the egg turns to scar tissue if the egg is not fertilized. However, in just a few months or so the entire ovary would be turned to scar tissue. Thankfully God made it so that the one place in a women's scars dissolve is in the ovaries.

2. Once an egg has been fertilized in the oviduct it will then move down to the side of the mother's womb where it will implant in the uteran wall. The egg is able to even get in the uterus because of the cillia of the oviduct, which beat in a metachronal rhythm. Again, another evidence of the designer.

3. We also gotta have the sticky cumulus cells that surround the egg in order for the egg to move through the oviduct, yet another step necessary for reproduction.

4. When the sperm meets the egg it has to be able to get through the cumulus, to this end the sperm secretes a special enzyme that was created to digest a path through the cumulus. Without having that particular enzyme we wouldn't get a baby.

5. The sperm also secrets a second enzyme in order to get the zona pellucida, what is the chance that that sperm had two special enzymes needed to penetrate the egg?

6. The egg's cell membrane knows to let the sperm in, yet another evidence of purposeful creation.

7. Once a sperm touches the egg's membrane, this starts an amazing response from the egg, known as the cortical reaction. This makes tiny granules break and secrete oval peroxidase (an enzyme). This enzyme surrounds the zona

pellucida and hardens it. Preventing the remaining sperm from penetrating the egg. Yet another coincidence according to evolutionary theory.

8. It is a miracle in and of itself that foreign tissue (aka the blastocyst) is able to burrow into the uteran wall without triggering an inflammatory response

9. The mother's blood clotting mechanism is suppressed within the placenta as it merges the sisal trophoblast with her blood vessels.

10. During birth, the placenta is also ejected. This causes twenty major arteries that are attached to the mother's womb to be severed. Thank Jesus that there are purse-string like sphincters that sinch up to prevent the mother from bleeding out.

11.Also during birth a special enzyme is made that partly dissolves the joints in the pelvis so that the bones (or ligaments) can stretch just the right amount to let the baby through.

12. At birth is the first time (in quantity) that the baby gets blood circulation in the lungs and liver, you could not gradually evolve that or else the baby wouldn't make it.

Origins-Fearfully and Wonderfully Made

**101. Homologous Structures evidence for a common design not descent.**

" Kent Hovind Creation Seminar 4, 1hr 8min"

**102. Bird Migration.**

Bird Migration is another thing that confounds evolutionary theory. If the universe was made by random chance without a creator, and if life ultimately has no purpose then how do we possibly explain that specific species of birds know how to migrate from one part of the earth to another. This is especially true for the babies on their first flight to wherever they are going. If they didn't follow their parents to a specific location before then how do they know where they are supposed to go. The blackpoll warbler and the black rumped petrel are just two examples of birds that I can give.

The Evolution Handbook by Vance Ferrell pg 205-206,715

## 103. The woodpeckers.

The woodpecker is another astounding piece of evidence against evolution. Why? You might ask. Well, I will just give two reasons. First of all the woodpecker's tongue is not like a normal tongue and I cannot even begin to imagine how it might've evolved the way that it is. Most animals tongues are located in their mouths as I'm sure that you are aware. The woodpecker however has a tongue wraps around the skull in a sort of spiral-like pattern. Secondly in addition to having barbs on it's tongue the woodpecker also secretes a glue-like substance on it's tongue to help capture tasty grubs. That's great in all, but if evolution was true how do we explain the woodpecker not dying when it glued it's beak shut! The glue and the glue solvent mechanism would've had to evolve simultaneously or the glue solvent would've had to evolved first which would require forethought, which is not something evolution can produce.

"Incredible Creatures that Defy Evolution Vol.1"

## 104. Termites and the protozoa in their guts.

Here is another symbiotic relationship that refutes evolution. Termites eat wood but cannot digest it, the termite relies on a little micro organism in their guts to break down the cellulose so it can get nutrients from it. The termite in return provides an environment for the protozoa to flourish.

https://newcreation.blog/mutualism-when-organisms-work-together/

## 105. World's Oldest Tree.

Estimated to be about 4600 years old (according to link) or around 4,855 years old (according to google), Methuselah a great basin bristlecone pine located in California is said to be the oldest tree by tree ring data. Hmm. If the earth is millions of years old, then why don't we find tree's at least 6 or 7 thousand years old. This seems more consistent with the biblical flood data than long age for the earth data. Also it is a fact that some tree's (including the Bristlecone pine) can grow 2 rings per year, under the right circumstances.

https://creation.com/evidence-for-multiple-ring-growth-per-year-in-bristlecone-
pines#:~:text=The%20ages%2C%20however%2C%20are%20based,than%20one%20ring%20per%20year.

## 106. Monarch Butterfly.

It "just so happens" that the majority of caterpillars are made to eat only one type of leaf. For example: The Monarch butterfly prefers to place it's egg on the milkweed plant for it's offspring to chow down on. You place the grub on the wrong leaf, it dies. How does the butterfly know which leaf to leave it's egg on? When the caterpillar gets to a certain point it goes into it's cocoon stage, it sort of dissolves into this liquid phase where it's genetic makeup is changed and it metamorphosis into a butterfly. How does this newly formed insect know how to fly? And how does it know what to do with it's proboscis for that matter?

"Incredible Creatures that defy Evolution 3"

## 107. Fruit flies.

Fruit flies are perhaps one of the greatest evidences against evolution. Why is that? This is because even though scientists have spent most of the 20th Century irradiating fruit flies, they still don't change into anything new. They remain fruit flies. Fruit flies were great for this experiment because they...

· Are cheap to feed

· It takes very little time for them to reach maturity 12 days, plus they have a short life span.

· The fruit flies also have many easily observed characteristics (such as the form of the wings, the eyes structure and color, just to name a few).

· And they only have a few chromosomes per cell.

If the fruit fly cannot change into another species after many many many generations, no matter what scientists try to do to it, why do we think random forces can cause such changes with much less mutations?

The Evolution Handbook by Vance Ferrell pg 36, 345-348

## 108. Chromosomes.

Just because an organism has more organism than another organism, this does not mean the one is more evolved that the other. A fruit fly has 8 Chromosomes but a house fly has 12, that's 50% more chromosomes for such a similar creature, it's the same kind of animal. The tomato has the same number of Chromosomes as the housefly but a four year old could tell that those two things are different.

The opossum, kidney bean, and Redwood tree all have the same number of chromosomes (22), yet it takes no genius to figure out that these three things didn't evolve from the former three things (the flies and the tomato). Nor are they similar just because they have the same amount of chromosomes. I am assuming evolutionist believe Humans are the most evolved creature in existence (well on earth anyway). Well how can that be if Chromosome count has any correlation with level of evolutionary development. Humans have 46 Chromosomes and chimps have 48, so are chimps more evolved than us? Is tobacco more evolved than us (48)? There are many examples of animals that have more Chromosomes than humans including the dog (78), turkey (82), and Goldfish (94) just to name a few. The second source I list here contains many more categories with chromosome counts, feel free to check out the book (the evolution handbook) for that information.

Kent Hovind Creation Seminar 4 2hr 8min

The Evolution Handbook by Vance Ferrell pg 712

## 109. Law of Kinds

What is the Law of Kinds? We see in the very first chapter of Genesis that animals bring forth after their kind. There is the dog kind, the cat kind, the horse kind, the pig kind, etc. The man-made taxonomy chart doesn't list kind in it's categories; it goes from the broadest being Domain to more and more specific,...kingdom to phylum to class, to order to family to genus to species to subspecies. I have included in this section a link that may prove helpful to the reader to distinguish what kind a certain animal is, or at least narrow it down a bit. Here we see a chart of several different types of animals, if you do some research you can easily find out from a google search that a dog and a jackal can reproduce but a dog and a fox cannot, therefore the dog kind would be located in the genus category for this particular example.

All observable evidence points to macro-evolution being false.

https://courses.lumenlearning.com/wm-biology1/chapter/reading-the-taxonomic-classification-system/

## 110. Toads.

How do Toads confound evolutionary theory? I am glad you asked. During the construction of New York's Eerie Canal (which occurred from 1817 to 1825), gravel, limestone, and drifts of clay were cut into. These rocks were dated to have been deposited during the Ordivician/Silurian period, according to evolution. These rocks are supposedly around 419 to 485 Million years old. When they removed solid limestone rocks from this area they encountered live toads encased in this limestone, some of which after being revived from their stupor (hibernation) survived for days. So now we have two scenarios...

A). Animals can live in a hibernation like state for hundreds of millions of years or...

B). Those rocks are not millions of years old.

Guess which one the evolutionist would pick.

Genesis week Season 2 episode 25

## 111. The complexity of the eye, and especially the trilobites eye.

According to evolutionary theory we came from a "simple cell" organism over 3 billion years ago. Over time life supposedly got bigger and more complex. How is it then? My dear readers that the supposedly ancient creature (the trilobite) has such a complex eye? The trilobite's eye is one of, if not the most complex eyes found in nature and yet it is one of the older or first evolved organisms.

" Kent Hovind Creation Seminar 4, 32min"

## 112. Lampsylis Mussel

Check the video for a full description, but basically this little Mussel has a lure that mimics a tiny minnow, this attracts the large mouth bass. The moment the bass begins to strike the Mussel shoots it's eggs into the bass's gills.

"Incredible Creatures That Defy Evolution 3"

## 113. Living Fossils.

1. **Trilobites**
2. **Graphtolite**
3. **Coelacanth**

" Kent Hovind Creation Seminar 4, 32min"

## 114. Bird Co-op.

One of the mantras of evolution is "survival of the fittest", it's the idea that only the strongest, smartest, fastest, or in other words the most fit of the species survive. Isn't it strange then that certain creatures of the animal kingdom instead of competing against each other for food, work together in co-operation for their meals. In this example I will list how birds search for their food.

· Woodpeckers dine off of the trunk and branch, drilling in.

· Nuthatches feed on the bark going toward the earth.

· Creepers work on the bark going upward.

· Kinglets eat off of smaller twigs as well as munching off of foliage

· While we see warblers eat off of the end of twigs and catch food from the air.

The Evolution Handbook by Vance Ferrell pg 838

## 115. Giraffe Part 1.

As I'm sure you are well aware the giraffe has a very long neck, this requires a very powerful heart to pump all that blood against gravity to the animals brain. Here is the problem for evolutionary theory. When the giraffe goes to get a drink of water, how come the giraffe doesn't blow his brains out when he lowers his head. You would think this would be the case since his powerful heart that was pumping against gravity, all of a sudden starts to pump with gravity. It turns out this animal has little spickets or you can call them valves going up his neck and they close, additionally the giraffe has a small sponge underneath his brain that absorbs that excess blood. Here is another scenario for the giraffe. Say he is getting a drink of water when a lion starts to approach. If he got up too fast wouldn't he pass out after a few steps due to lack of oxygen to the brain? No, because the spickets open back up when he gets up and that sponge that held that oxygenated blood releases it into the brain.

"Incredible Creatures that Defy Evolution...1"

## 116. The Heart/Circulatory system.

The Circulatory system poses yet another problem for evolutionary theory. You see, evolutionists tell us that we evolved from fish many, many millions of years ago. hear in lies a problem. Fish hearts have two chambers, amphibians three, and humans have four chambered hearts. Just a small tear in your Circulatory system and you might bleed to death. So the problem is, how can our supposed fish ancestors have made all these massive or multiple changes to their circulatory system without them dying in the process. I would imagine You would have veins going to nowhere causing the animal to bleed to death internally, or you would have organs that were beginning to forming but not yet connected to the veins, in which case the organ cannot form without receiving nutrients from the veins. For more information, please check the link below.

"Genesis Week Season 3 Episode 3."

## 117. Beans.

Now how do beans disprove evolutionary theory? That is a good question. Unless an evolutionists believes in punctuated equilibrium (see point 147), then he probably believes in slow, gradual changes that result in a new organism. We wouldn't expect there to be limits or constraints to this change. This is where beans come in. Round about the time of 1900 Mr. Wilhelm Ludwig Johannsen bred extremely large and small beans. Instead of continuously getting larger and larger or smaller and smaller beans over multiple generations, Johannsen discovered that after a few generations the beans reached a certain limit in size and were unable to go any further in that direction.

The Evolution Handbook by Vance Ferrell pg 389

## 118. Nodoaur Mummy

Over and over again we see in children's books that dinosaurs lived "millions of years ago". The mantra goes like this, "Dinosaurs died out 65 million years ago". Well then we need an explanation then. If dinosaurs haven't lived for many millions of years how do we explain this Nodosaur mummy? This is **Not** a fossil (which would be a stone cast of was once organic matter). It has skin scales with kerogen (Kerogen is a secondary molecular structure that is formed when proteins break down and mix under the earth). This specimen also contains the original pigments. There is no good explanation of how these pigments and proteins can last millions of years.

https://www.icr.org/article/spectacular-dinosaur-has-skin-horn

## 119. Cuttlefish.

I will not go into too much depth here about the cuttlefish (check out the video reference below to learn more; free on youtube). I will say however that the cuttlefish ability to camouflage itself into it's environment by not just changing the color of it's skin but the appearance of texture as well make this creature an affront to the notion that it could have evolved by chance. Additionally the ability to stun it's prey with dazzling strobe lights, pretty cool.

"Incredible Creatures that defy Evolution 3"

## 120.Beneficial mutation?

Evolution presupposes countless beneficial mutations in order for it's modal to work. If evolution were in fact true, then we should expect evolutionists to be able to give us many examples of a beneficial mutation. Probably the most popular example of a beneficial mutation is that of sickle cell anemia. This is referred to a "beneficial mutation" because those who have it are somewhat resistant to malaria. But the question is, do the pro's really outweigh the con's? 25% of the children of carriers will most likely die of the anemia, and another 25% can contract malaria. What's more the gene that causes sickle-cell anemia tends to die out in areas where malaria is not a big problem. In my opinion calling sickle cell anemia a beneficial mutation is like saying a limbless guy is the result of a beneficial mutation because he cannot get handcuffed.

The Evolution Handbook by Vance Ferrell pg 343-344

## 121. Selective Breeding (Chihuahua)

A major doctrine of evolution is upward and onward change. That animals over time are getting bigger or smarter, or stronger; and that they are gradually changing into something different than their ancestors. Natural Selection is what we call a misnomer. Natural selection should be titled "random occurrences". This is because selection is the result of careful planning and a mind. Evolution is said to be an unguided process. What's more is that if selective breeding cannot break the kind barrier, how can unguided random processes do it? Although selective breeding may enhance a specific trait of an organism, it always makes the whole of the organism less hardy. A prime example of this is the chihuahua. They are bred to be tiny little creatures. But for the most part they are a neurotic and nervous mess. They would not last long in the wild.

The Evolution Handbook by Vance Ferrell pg 393, "Hovind seminar 4, 47min"

## 122. Symbiotic Relationships: Milopina Bee.

The vanilla plant is an orchid native to Mexico, that only blooms one morning for a few hours and then wilts. This plant is where we get the vanilla bean from. The Milopina bee apparently is the only insect that know how to enter this flower to extract the pollen. Without this bee the plant goes extinct. What would be the odds that both these organism arrived by chance on earth at the same time, in the same location, and how does the bee know what to do to get into the orchid without a designer. The truth is this would be impossible. The creation testifies the brilliance of the Creator.

"Incredible Creatures that defy Evolution 3"

## 123. Human Brain.

The human brain is another enigma for evolutionary scientist. According to evolutionary theory (or at least Darwin's book 'origin of species' natural selection should only make mankind just be smart enough to be able to survive in his environment. The thing is mankind's brain is way more advanced than he really needs. If man only needed to be just smart enough to kill a few small animals and gather some plants why develop a brain that is capable of building complex space shuttles, high tech cell phones, and complicated music?

Contrast this with the brain of apes and we can see a humongous difference. Apes can hardly use tools, the reason why apes seem smart is because the apes keepers just assume that they are.

The Evolution Handbook by Vance Ferrell pg 581

## 124. Bio-mimicry. Riblets.

Bio-mimicry is when scientists copy the design they find in nature to make or improve products for customers. A great example of this are the riblets found on fast moving fish. These riblets are triangular shaped grooves on the fishes skin that help reduce the friction/drag of the water. This design was used on airplanes and has saved businesses massive amounts of money in fuel costs.

The Evolution Handbook by Vance Ferrell pg 66

## 125. Mallee Bird.

The Mallee Bird makes it's home in the Australian desert. About the time of May or Junes, this bird uses it's claws to make a hole about 3 feet in depth and 6 feet in length. Afterwards it is filled with vegetation and when it rains the ditch is heated to over 100 degrees Fahrenheit. When the temperature is down to 92 degrees the mallee bird calls for it's mate and they, well, mate. The female lays one egg a day for a month and then takes off.

Then comes the really amazing part the male bird covers the eggs with sand and continuously check on them to make sure they stay at the right temperature 92 degrees for seven weeks. If it heats up during the day he takes off sand, if it cools at evening he adds more sand. The temperature cannot vary even one degree. Once the chicks are hatched they climb outta the sand and are actually able to fly right away. When they get older they repeat the process.

The Evolution Handbook by Vance Ferrell pg 66

## 126. Portrait Frog.

This is one frog that evolution will never be able to explain. It is another clear example of the teleological argument (argument from design). The portrait frog (also called the south American false eyed frog) has the uncanny ability to change it's colors on it's back. During the dire crises of encountering a predator the portrait frog hops around so that it's back is facing toward the would be predator. Then an amazing thing happens, the colors on the frogs back change to resemble that of a giant rat head. Normally the frog sits with it's head up and bottom low, but in this instance the frog's head is down and it's bottom is up. During this confrontation the frog has tucked his back legs underneath the spots that look like eyes so that the legs look like a moving mouth. The part of the portrait frog that once was it's tale during the tadpole phase, now appears as a nose in just the right location. What's more is that the frogs back feet appear as intimidating long claws to the predator.

Evolutionary theory cannot explain this. The frog could never draw something on the ground in front of it like what we see on it's back. Any sane person would say that the frog would lack the intelligence to do so. How much harder would it be to make something like this on it's back apart from a creator God. Scientists would tell you that it wasn't the frog intelligence that caused the pattern on it's back but it's genetic code, and they'd be right. But how can blind, purposeless processes produce such an amazing pattern that resembles another creature? It takes way more faith to believe this happened by accident than to attribute it to God.

The Evolution Handbook by Vance Ferrell pg 897

## 127. Ichneumon Wasp

What is so remarkable about the Ichneumon wasp that it defies evolution? This tiny creature uses it's small thin antenna to drill straight down the bark of an oak tree to reach the larva of a specific beetle's larva. It then leaves it's egg on the larva so that when the wasp's eggs hatch they will have food to eat. The baby wasp then climb out of that tiny hole and repeat the same pattern without any instruction from their mother!

The Evolution Handbook by Vance Ferrell pg 838

## 128. Desert rat

Well what's so cool about this creature that warrants it's spot on this list of biological proofs against evolution. It turns out that the desert rats that make their abode in the western United States have the ability to make their own water. How do these small rodents do this? They actually mix the hydrogen which they get from ingesting dry seeds with the oxygen from the air they breath and presto, water!

The Evolution Handbook by Vance Ferrell pg 158

## 129. Bird Designs.

Have you ever thought about how well birds are suited for their environment? There are many kinds of birds in the world and they come in many forms and colors. Birds tend to be more colorful out in the field (because they can see danger afar off), but forest birds are more camouflaged to their surroundings. Ducks have webbed feet for swimming and a beak suited for gathering food from the mud and grass, water fowl also have thick, oily skin and thick feathers as well for repelling water. Wading birds have long legs, perching birds possess slender legs and tiny feet. There are short beak seed eaters, long beak woodpeckers, and slender beaks for the insect eaters out there. Birds have rib muscles like us but they also possess flying muscles, when the bird goes to flight the rib muscles lock up so the flying muscles can work properly. So how does the animal then breathe? The wing muscles take over the respiratory function and cause the air sacs to contract/expand.

The Evolution Handbook by Vance Ferrell pg 964

## 130. Tiger Moth.

Here is an evolutionary scenario for you. After millions of years of being eaten by the bat, the tiger moth thinks to himself, hey I need to do something about this echo location, I keep getting eaten! It is then that the tiger moth slowly over millions of years starts the complicated process of learning how to build radar frequency jamming equipment into his genes. Of coarse this is a ridiculous scenario. Yet we see that the Tiger Moth does have this equipment. Now is it more likely that God put that in his gene code or did the moth slowly evolve it over an incredible long time?

The Evolution Handbook by Vance Ferrell pg 855

## 131. Wondernet

Despite human beings being the supposed pinnacle species of evolution they would not last long in freezing temperatures. Especially when submerged in freezing water. Then how do seals and whales swim all the time in the freezing arctic waters? This is because of the animals wondernet. A wondernet is the seals/whales circulatory system which is quite different than our own. The Seal/ Whale's veins are designed to withstand this freezing water. Not evolved. For if it had to evolve this way, the animal would die way before it had the chance. The blood from one vein travels in the opposing direction of the blood of the nearby adjacent vein, in this manner warm blood passes on it's heat to colder blood. This is how these creatures manage to stay warm.

The Evolution Handbook by Vance Ferrell pg 371

## 132. Black rumped Petrel.

When we see how things in nature work in a just so manner, it compels us to acknowledge our Creator. Such is the case of the black rumed petrel. When it's time to nest, this bird (from wherever is at in the wide pacific ocean) travels to the Hawaiian Islands. From there it goes to the tallest mountain (Haleakala), which is an extinct volcano. It is said that this volcano has the widest crater on earth. This mountain is ten thousand feet high. Therefore the nest is higher than any other ocean bird! The incubation period for this egg is the longest for any bird (55 days). At such a high altitude the oxygen is very thin, this could be dangerous for the baby chick because the egg needs to absorb oxygen through it's tiny holes and emit vapor as well. The Black rumped Petrel egg has the least amount of holes of any bird, however because eggs absorb oxygen faster at high altitudes this works out just fine. The baby chick is fed by it's mom and paw for four months, since the food is so far away in the ocean, the Petrel makes a rich oil in their bodies in which to feed their young.

The Evolution Handbook by Vance Ferrell pg 613

## 133. Firefly

Evolutionist will tell you that fireflies evolved by chance over millions of years, how is it then that scientists cannot create with their super smart brains what supposed mindless random accidents done millions of years ago? I am referring to the fireflies light! 90% of the energy used to make light does exactly that (make light) for the firefly; whereas when scientists try to make light, over 90% of it is used as heat.

The Evolution Handbook by Vance Ferrell pg 371

## 134. Bee Anatomy

According to evolution bee's evolved from wasp around 120 million years ago. If that was so should we expect the following 19 features from a creature that is supposedly way less evolved than we are.

1. Antennae for touching and smelling. 2. Front legs that have grooves on them to clean said Antennae.

3. Tube shaped proboscis that curls under head when not in use. 4. Compound eyes used for navigation.

5. Two mandibles that is capable of forming wax, as well as holding and crushing things. 6. Beeswax producing glands. 7. Glue-making glands. 8. Honey tanks for nectar. 9. Enzymes in honey take to make honey. 10. Head glands make royal jelly. 11. Segmented legs that go in any direction. 12. Claws on every foot for the purpose of clinging to flowers. 13. Pollen collecting hairs all over body.

14. Back legs have little pollen baskets. 15. Other structures that are used for pollen collection. 16. Poisonous stinger for defense. 17. Pollen packing spurs. 18. Front wings hook to back wings for improved flying power. 19. A wealth of knowledge on how to guard and make the hive, show others where the flowers are located, nurse babies, aid the queen, collect materials, and much much more.

The Evolution Handbook by Vance Ferrell pg 895-96

## 135. Air sailing spiders

Did you know newborn baby spiders can sail? Air sail that is. Soon after they are born they send out a nine foot thread with their little spinneret and off they go! They they climb to the middle without falling (that in and of itself is a mystery) and tuck the thread here and there, which creates a rudder for which it can steer. How can this tiny creature with a microscopic brain that is barely out of it's egg know to do this? Easy for the Creationist to say, it was Jesus who programmed this into the creature.

The Evolution Handbook by Vance Ferrell pg 715

## 136. Seabirds.

So what is so spectacular about seabirds? Well seabirds have a built in Desaltation system inside their heads that cause them to discharge a highly concentrated saline solution from their beak. So how do evolutionists explain how the birds could've evolved this, since drinking saltwater can be fatal to an organism. Did all the birds die for millions of years before evolving this trait? Dead creatures don't evolve.

The Evolution Handbook by Vance Ferrell pg 830

## 137. Sunbird and Mistletoe.

The Mistletoe would've gone extinct without the Sunbird. This is because the mistletoe doesn't open up without the sunbird coaxing it with it's long beak. Once the bird inserts it's beak into the mistletoe it is shot with pollen which the bird carries to the next flower, continuing the species.

The Evolution Handbook by Vance Ferrell pg 742

## 138. Rat teeth.

This is an example of Bio-mimicry, which is studying nature to in order to solve complicated human problems. When wanting to design a self sharping saw blade, engineers at General Electric studied the teeth of rats. They discovered that it could not have been done as efficiently at their teeth. Rat's top teeth go behind their bottom teeth at such an angle that it sharpens them. Though not in my reference page below I'd like to add that rat's teeth continuously grow throughout their lifetime.

The Evolution Handbook by Vance Ferrell pg 281

## 139. Mutations friend or foe to evolution.

Although Neo-Darwinist like to imagine that Mutation+Natural Selection = New Animals, this simply defies logic. This is for at least four reasons. For one, mutations are very rare. We are talking of such odds as 1 in 10,000 to 1 in a million, per gene, per generation (in higher organisms). For evolution to be true we need millions of changes.

Secondly, mutations are random, it has never been observed to create a new kind of animal, we know this by inducing hundreds of thousands of mutations on fruit flies and other tiny critters. Thirdly, mutation does not help the organism, well over 99% of mutations damage the organism. Lastly, mutations are always and/or almost always harmful or lethal to the organism. My source list 28 reasons <u>why mutations cannot produce a cross species change</u>, I will list some of them, but I highly recommend getting the book to learn more.

1. There are no recordings of a truly beneficial mutation. (One that was permanently passed down for generations and that wasn't just a reshuffling of latent characteristics).

2. Doesn't it seem quite hypocritical for evolutionists to claim mutations are helpful to the species, yet won't let their own bodies be bombarded by X-rays for minutes at a time. There is a reason why mutagenic chemicals are disposed of so carefully.

3. Interconnected wiring-Each gene affects multiple characteristics, and each characteristic are affected by numerous genes.

4. The Randomness of mutations would negate any positive change made to an organism. What good are eyes on your armpits, or feathers on a birds feet?

5. According to the theory of Syntropy organisms wouldn't last very long at all unless they were complete and had all of their functions working perfectly or close to perfect.

6. It is mathematically impossible for a DNA strand over 84 nucleotides to be the result of random mutations. At that level the probability is $480 \times 10^{50.}$ According to mathematicians $10^{50}$ has no chance of happening. Even simple cell bacteria have 3,000,000 nucleotides. We have no known species that could have arisen by chance.

7. Evolutionary Theory supposes ever-increasing complexity, upward and onward to better organisms. Expecting radiation or mutagens to create more highly evolved creatures is sorta like taking a rifle and blasting away at an automobile and expecting to getting a cooler looking, better performing car.

8. In order for evolution to work we need New structured Information added to the organism, random processes cannot produce this. I would like to add this tidbit that is reinforced elsewhere in this book (point 254) Information always comes from a mind.

9. Evolution requires new organs.

10. Unlike what the Comic series X-men portray mutant kids are never as strong as their unmutated/less mutated parents, according to geneticists.

Finally I will say this last remark. According to the research using 5-bromouracil to increase mutations by a factor of ten thousand. Scientists finally discovered the truth about mutations, they are not 99% harmful to the organism, mutations are 100% harmful to the organism. Every single instance of mutation damaged the organism. DNA just can't cope with anything but the slightest amount of change, more than that and it will greatly weaken the creature.

The Evolution Handbook by Vance Ferrell pg 320-342, 352

## 140. Survival of the Fittest

Although taught in evolutionary circles, the doctrine of survival of the fittest is actually a friend of creationists. How is that? It's quite simple. The teaching of survival of the fittest claims that the weak of the species gets culled out of the gene pool. Mutations (the supposed hero of evolution) from all observable evidence, actually weakens organisms (as described in the previous point), these are the organisms that are culled out of the species; keeping the species very stable. So it turns out that Survival of the fittest does the opposite of what evolution requires, which would have one kind of animal turning into another.

The Evolution Handbook by Vance Ferrell pg 287

## 141. Termite Towers.

Are just incredible! Termites are blind, tiny, and seemingly not very intelligent, how can they be with a brain that small. Yet the termites are able to build some of the most fancy, and complex structures in the animal kingdom (my words, not the sources'). Termite towers are massive in comparison to the workers (some rising to an astounding 20 feet)! The complexity of the structure is an excellent example of an intelligent creator, for how can these blind little termites know how to do the following.

· They build in relation to magnetic north

· They structure it so it gets heat in the morning and later in the afternoon, but less heat around noon.

· The roofs are conical with sloping sides ranging from smaller at the top to bigger as you go down as well as outward projecting eaves to bounce rain away from the base of the tower.

· There are nurseries, storerooms for food, and fungus garden as well.

· There is a special area for the king and queen.

· This place even has air conditioning.

The Evolution Handbook by Vance Ferrell pg 316

## 142. Back and forth change

According to the Evolutionary theory animals can only evolve in one direction. What do I mean by that. I mean a fish can turn into an amphibian, but an amphibian cannot turn into a fish. A dinosaur can turn into a bird, but a bird cannot turn into a dinosaur. The point that I am trying to convey is that according to evolution an animal cannot return to what it once was (presuming evolution was true). Is there evidence for this assumption though. None, whatsoever!

In fact we have plenty of evidence to the contrary. We see that animals only reproduce after their kind and that sub-species changes can go back and forth. White peppered moths can give rise to dark and vice versa. Pigeons can give rise to different colored/shaped pigeons and back again.

The Evolution Handbook by Vance Ferrell pg s 294

## 143. Animals with other animal patterns (Mimicry) and camouflage.

Paintings and Information require Intelligence. The theory of evolution leaves out the intelligence from biological changes. Evolution is supposed to be a blind, purposeless, unguided process. Then how do we explain how an animal or insect develops these complex patterns that resemble other animals or plants? It's quite a stretch to say evolution did it don't you think? Below are five examples of Mimicry and Five examples of camouflage, forgive me for giving secular sources; I was aiming for providing nice photos for demonstration.

Animals that copy other animals

1. Hornets and Hornet moths.
2. The ladybug and ladybug spiders.
3. Jumping Spiders and Metalmark Moths.
4. Hawk Moth Caterpillars and Pit Vipers.
5. False Cleaner fish and Blue streaked cleaner wrasses.

Animals that blend into plants or the environment

· Screech owl resembles the tree.

· Pink Pygmy seahorse blends in the coral reef.

· Mountain hare matches snow.

· Dead Leaf butterfly, and well...leaves.

· Leaf-tailed Gecko and Bark.

https://listverse.com/2019/08/28/10-extraordinary-cases-of-biological-mimicry/

https://mymodernmet.com/camouflage-animals/

## 144. The Eye.

"To suppose that the eye with all its inimitable contrivances for adjusting the focus to different distances, for admitting different amounts of light, and for the correction of spherical and chromatic aberration, could have been formed by natural selection, seems, I freely confess, absurd in the highest degree."

Charles Darwin, The Origin of Species, J. M. Dent & Sons Ltd, London, 1971, p. 167.

https://www.creationworldview.org/evolution-is-a-religion

## 145. Distinct Species shouldn't exist

Another big problem for evolution is the mere existence of species...bear with me.

Evolution teaches that we are constantly evolving (changing), into something else. Why then are there distinct species. It is ridiculous to believe that we are evolving in unison and following the same evolutionary course. Evolution is supposed to be caused by random mutations happening in the individual which are then passed on to the children. If evolution was the truth shouldn't we expect to see people with octopus arms, or bears with fish scales or something. Like each individual person should have some unique quirky or dangerous thing about them. This isn't was we see at all though. Cheetah's look like cheetahs, birds look like birds, Apes look Identical to each other, once of them hasn't sprouted wings.

The Evolution Handbook by Vance Ferrell pg 304

# Frauds and Hoaxes (Biology)

## 146. Peppered Moth Fraud.

The peppered moth has recently and may still be touted as evidence for evolution in some school textbooks. They supposedly studied some moths on trees and the story goes that before they burned coal in the factories the trees contained 95% white moths and 5% black ones (because the birds preferred black moths. But afterward when the trees turned black there were 95% black peppered moths and 5% white, because the black moths were camouflaged. The problem with this story is that it was a fake. They glued 2 dead moths on the tree in order to take the picture. This is supposedly the best evidence for natural selection and was given ten times more than any other evidence in England. Even if this example wasn't a hoax this still wouldn't prove evolution, even creationist agree that there is considerable variation within kinds.

The Evolution Handbook by Vance Ferrell pg 289, Hovind Seminar 4, 1hr 6min

## 147. Punctuated Equilibrium

Richard Goldschmidt (geneticist) and Stephen Gould (Paleontologist) could not find evidence of slow-gradual evolution in their respective fields so they came up with two theories which basically the same thing. Instead of slow gradual changes over time causing one species to change into another; it happens all at once. In other words in one generation a brand new distinct species emerged. It would be akin to a chicken hatching from an alligator egg (my example). The amount of blind faith to believe such nonsense is more than i could ever hope to muster. There are multiple enormous problems with this theory. Here are four.

- Would require multiple millions of beneficial changes in a single generation, we (arguably) observe no beneficial mutations today.

- Additionally this would happen twice around the same time (within a few years).

- It would have to produce a male and a female.

- They would have to live within a few miles of each other and survive long cnough to matc.

The Evolution Handbook by Vance Ferrell pg 356-57

## 148. Vestigial Organs Pt.1

A vestigial organ is an organ that is a supposed leftover from our ancient ancestors and has since lost it's function. Some evolutionists may say that the supposed vestigial organ had a greater function in the distant past but now little or almost no function. This is a belief system for which there is _no evidence_. For one, the only tangible evidence of what our ancestors were like are a relatively small amount of bones, and secondly, how can bones tell us in great detail what function a specific organ did when the creature was alive? I would love an explanation for that.

Anyway there are no organs in the human body that are actually vestigial in the sense that they serve no function period! If you search the internet you may come across the statement that there are more than 100 vestigial organs in the human body. This statement is derived from German Anatomist Weidersheim's 1895 work. For the most part these supposed examples of vestigial organs were from small muscles or tiny variations in bone. A lot of these muscles were called vestigial because they only contributed only a small part to the total muscle force or made no apparent contribution. However if a muscle isn't being used it will shrink; as documented in outer space weightless situations. We know understand that a lot of those small, short muscles are used to make adjustment in the larger muscles. They are also used for proprioception (this allows for quick as well as accurate limb positioning). This is why felines seldom fall on their back. According to David Menton (Anatomist), small muscles are used to serve a sensory function rather than a motor function.

**Below are some resources for further study.**

https://answersingenesis.org/human-body/vestigial-organs/vestigial-organs-vanishing-argument/

" Kent Hovind Creation Seminar 4, 1hr 37min"

https://genesisapologetics.com/wp-content/uploads/2014/07/Chapter-9-Vestigial-Structures-from-Biddle_Creation_v_Evolution_7_1_2014-for-web.pdf

**149. Haeckel's Embryo's Fraud.**

Haeckel's Embryo's have been used to support the theory that all organisms share a common ancestor. The images portray a human, a rabbit, a hog, fish, salamander, chicken, tortoise, and calf side by side and look remarkably similar....here's why:

1. The embryo's earlier stages were faked, some were mere copies of others, while others were made to appear more similar than reality portrays.
2. The examples were cherry picked to stack the deck in his favor.
3. He actually omitted the earliest stages and claimed the mid-stage as the earliest stage. The earliest stages look quite different from each other.

The Case for a Creator by Lee Strobel 47-50

## 150. Gill slits Fraud.

Ontogeny recapitulates phylogeny is a complicated way of saying that babies in the womb repeat their evolutionary history by going through the mature stages of their ancestors as they form. Gill slits is probably the most common example cited by Evolutionists. Babies do not form gill slits in the womb, they only appear as gill slits in the superficial sense, it is only imagined. To make a comparison it's like interpreting the bible using eisegesis instead of exegesis.

"The Case for a Creator by Lee Strobel" 51-52

## 151. Vestigial Organs Pt 2

For this short section I will list seven common examples of vestigial structures in humans, examples that evolutionists recklessly claim have no apparent function. If an evolutionists will claim that these organs have a reduced function from what it used to have, it is on them to provide evidence for such assertions. They will not be able to, as their is no tangible evidence of what at organ was like supposedly "200,000 years ago".

1. **The Coccyx**-Serves as an attachment site for muscle fibers and tissues that support various lower body organs.
2. **The Tonsils**- Helpful to the immune system, particularly in youth.
3. **Appendix**- Replenishes the bowels with good bacteria.

4. **The Thyroid**- Produces T3 and T4 which regulates metabolism and calcitonin.
5. **The Thymus**- Needed for the development of T-cells, also plays a key part in aiding with autoimmune issues.
6. **The Pineal Gland**-produces sleepy hormone Melatonin, this organ is also used in timing of puberty.
7. **Nictitating Membrane**-Aids in eye rotation and helps protect the eye from foreign material and also help remove foreign matter.

" Kent Hovind Creation Seminar 4, 1hr 32"

https://genesisapologetics.com/wp-content/uploads/2014/07/Chapter-9-Vestigial-Structures-from-Biddle_Creation_v_Evolution_7_1_2014-for-web.pdf

## 152. Lamarkism Error

Lamarkism is the theory of inheritance of acquired characteristics. It was dis-proven by August Weismann a little while before 1900. August cut the tails off of 19 rat generations. The babies always had their tails. The point is we don't acquire new organs from our ancestors, or lose them if they did. Circumcision is another proof Lamarkism is wrong. Babies do not come out of the womb circumcised. Another example I will give, one that is not in my reference, is that there is a long neck tribe of people (the Kayan); there women wear rings around their necks to elongate them, their young girls have to repeat the process, they are not born with unusually long necks.

The Evolution Handbook by Vance Ferrell pg 297

## 153. Horse Evolution fraud

A long time ago (way back in the 1800's), a guy name Othniel Marsh claimed to have discovered thirty different types of horse fossils in Nebraska as well as in Wyoming. When he reconstructed and organized these fossils they were then displayed at Yale. The horse series has the small creatures on the left while bigger ones to the right. There are over a dozen flaws in this horse series (for the complete list please see the source provided). Here I will list just five.

· The specimens to not come from the "right sequence" of lower to upper strata. Sometimes we have the most minuscule horse at the top layer.

· There are no transitional forms between specimens.

· The three-toed "more ancient" specimen is found in strata above the one toed "more recent" specimen.

· The smallest "horse" is really a rock badger!

· Horse size does not prove evolution, horses today vary wildly in size. Some horses are smaller than dogs, others are humongous.

The Evolution Handbook by Vance Ferrell pg 745-50

## 154. Giraffe Part 2

The second thing that I'd like to share about the giraffe that bucks against evolutionary theory is the tale of how the giraffe got it's neck. The story goes like this. Because of "survival of the fittest" the giraffes decided they would only survive if they could reach the leaves at the top of the tree's but at that time they supposedly had much shorter necks. Over time the giraffes stretched their necks farther and farther, ever higher and higher to reaches the tallest hanging leaves. So it is assumed that the ancestors of giraffes grew longer and longer bones in their legs and neck over millions of years. Three very BIG problems with this story.

1. This assumes all the short necked giraffes died of starvation because they refused to bend their neck down and eat the lush vegetation before them. They already had to bend over to drink. Other shorter animals were already eating ground level grasses.
2. The females would have died off anyway because they are two to three feet shorter than males.
3. There are no transitional fossils between giraffes and their supposed shorter necked ancestors.

The Evolution Handbook by Vance Ferrell pg 843-846

**155. Whale of a Tale.**

According to evolutionary theory not only did fish eventually turn into mammals over millions of years but apparently some sheep or deer or cow got homesick and decided to return the sea. So slowly over hundreds of thousands or even millions of years of wading on the shore of the ocean, learning to catch fish and slowly evolving the body parts and processes necessary to survive in the sea; such as moving it's nose to the top of it's head slowly over hundreds of generations, and developing the equipment to pump it's milk into it's calf's mouth to counteract the water pressure. The calf would also have to learn voluntary breathing and how to swim on it's back to avoid drowning until it evolved the blowhole in the right spot. Enough of these shenanigans, there is no possible way a cow, deer, or sheep could slowly evolve to live in the ocean, they would drown in the first generation, just like any other land creature would.

The Evolution Handbook by Vance Ferrell pg 852-854

**156. Whales Pt 2.**

Evolutionists love to point out the two little bones referred to as the whales pelvis, and say something like, Aha! Proof that whales used to have legs and came from a land dwelling ancestor. Evolutionists claim they are vestigial and yet admit they aid in reproduction. I thought vestigial means something that has no function.

" Kent Hovind Creation Seminar 4, 1hr 33min"

# How Biology Bucks against Old Ages

**157. Dino Soft Tissue.**

Example 1: When Mary Schweitzer and her group found soft tissue (red blood vessels and cells, as well as collagen) in a Tyrannosaurus femur back in 2004.

Example 2: Soft tissue in the rib and horn of a triceratops fossil from the Hell creek formation.

https://answersingenesis.org/dinosaurs/bones/dinosaur-soft-tissue-evolutionists-problem/?gad_source=5&gclid=EAIaIQobChMIxr3c7ISahQMVaWtHAR0CAAHREAAYASAAEgJwo_D_BwE

Toad you so! This is Genesis Week, episode 25, season 2 with Wazooloo/aka Ian Juby

## 158. Mollusk and Snails.

Long ages are inseparably linked to evolution, since we do not observe macro-evolution occurring today evolutionist assume that it happened in the distant past. For this they need deep time. Enter dating methods. Scientists use various dating methods to determine how old a sample is. However these dating methods are based on faulty assumptions which can really throw off the readings.

Two examples of faulty Dating method readings include C-14 dating of living organisms (namely Mollusk and snails).

- The ***living*** snails were Dated to be 27000 years old.

- The mollusk 2300 year old.

These errors were due to a process called the reservoir effect which introduced "dead carbon" into the groundwater. All the details for this process are found in the link below.

https://www.evolutionisamyth.com/dating-methods/problems-with-carbon-14-dating-method/

## 159. Seal.

So in what way would we expect the seal to favor Christianity over evolution? In this example the seal undermines the Carbon dating methods. When a freshly killed seals was carbon dated, we should expect the instruments to give us an age of 0 years old. That's not what happened, when seals were carbon dated they gave an age of 1300 years old.

https://apologeticspress.org/limitations-of-carbon-dating-5626/

# Other Interesting Facts

## 160. Lambs blood cures Snake venom.

Do I really need to go into great detail here? All veteran Christians should know what the Lamb and Snake represent. It is the Lamb's (Jesus') blood that washes us of our sin and is the cure for eternal death brought on by the Snake (Satan's) Lies and our sin.

Joh 1:29 The next day John seeth Jesus coming unto him, and saith, Behold the Lamb of God, which taketh away the sin of the world.

Col 1:14 In whom we have redemption through his blood, *even* the forgiveness of sins:

Gen 3:1 Now the serpent was more subtil than any beast of the field which the LORD God had made. And he said unto the woman, Yea, hath God said, Ye shall not eat of every tree of the garden?

https://www.quora.com/Does-sheep-blood-have-anti-venom

## 161. Female Mosquito.

The existence of animals that seem to be designed as predatory posses a problem for the Christian worldview because the bible tells us that their was no death till Adam sinned and that God gave the animals plants (green herb) to eat. What do we do with the animals that are presented as examples of predatory or parasitic. In this and the next point I will answer just that. It would seem cruel for God to create a bug that bites and feeds off of human. Thankfully there is ample evidence that not all mosquitoes drink blood indicating the possibility that the first mosquito didn't necessarily drink blood either.

Many mosquitoes need a blood meal in order to lay eggs. However those with the trait of autogeny can produce eggs without blood. This trait to not need blood for egg production is quite common in the culux genus of mosquito. The mere presence of this type of mosquito shows that the literal events of Genesis could have easily have taken place just as stated in the bible. Again, sorry for the secular link.

https://www.ncbi.nlm.nih.gov/pmc/articles/PMC4385763/

## 162. Lions.

How do lions refute evolutionary theory? Well, one example is little tyke. This female lion (a lioness) refused to eat meat. She was fed a diet of cooked grain, raw eggs, and milk. What's interesting is that the lioness would eat the tall, luscious, grass of the field, I (the author) can personally attest to house-cats eating my wife's indoor plants.

Isa 11:6 The wolf also shall dwell with the lamb, and the leopard shall lie down with the kid; and the calf and the young lion and the fatling together; and a little child shall lead them.

Isa 11:7 And the cow and the bear shall feed; their young ones shall lie down together: and the lion shall eat straw like the ox.

Isa 65:25 The wolf and the lamb shall feed together, and the lion shall eat straw like the bullock: and dust *shall be* the serpent's meat. They shall not hurt nor destroy in all my holy mountain, saith the LORD.

https://creation.com/the-lion-that-wouldnt-eat-meat

## 163. The donkey has a literal cross shape on it's back.

Zec 9:9 Rejoice greatly, O daughter of Zion; shout, O daughter of Jerusalem: behold, thy King cometh unto thee: he *is* just, and having salvation; lowly, and riding upon an ass, and upon a colt the foal of an ass.

## 164. Noah's ark.

One of the misconceptions people have had in the past (and perhaps some still do today) was that there would be no way Noah's ark could house all those animals. This false belief is rooted in two major lies. The first one is that Noah's ark was some tiny little boat comparable to a vessel not much larger than a school bus. This is probably do to Christian authors and programs that depict the ark in this way with cartoon photo's. The second is people see all the species of animals today and try and imagine them all trying to be squeezed inside said boat. The truth is Noah's ark was huge, a life-size replica has been made in Kentucky by a Christian named Ken Ham and his crew. Secondly Noah's ark housed all the necessary kinds (not species). There is only one kind of dog but many species, same with horses, Cats, etc. So Noah didn't take millions of different kinds of animals on the ark, hence eliminating the problem.

## 165. Law of Bio-genesis.

The law of bio-genesis states that life can only come from life. The foundational "truth" of evolution is that all life on earth came from non-living chemicals in earth's early oceans. These are contradictory and only one can be true. Scientific laws cannot be violated, that is what makes them laws. According to the truth of God's word we were created by the living God. This means the scientific law of bio-genesis was not violated.

Psa 84:2 My soul longeth, yea, even fainteth for the courts of the LORD: my heart and my flesh crieth out for the living God.

# Ch. 4 History is about His-story: Archaeological / Historical Evidence

## Archaeology & a Young Earth

**166. Luke Chapter 3.**

The gospel of Luke gives a genealogy that traces the lineage of Mary's husband Joseph all the way back to Adam. This alone is powerful evidence of a young earth. As Jesus states that he that made them at the beginning made them male and female.

**167. Cambodian Temple Stegosaur Petrogliph.**

Type this phrase in google, please. If you have any doubt that dinosaurs roamed the earth at the same time as man. Any open-minded person would come to the conclusion that this carving is of a stegosaurus. I list the following link that contains many pictures of the Cambodian temple and/or the petrigliph. It also answers a few objections from those who would doubt this is depicting a stegosaurus.

https://answersresearchjournal.org/stegosaur-engravings-at-ta-prohm/#:~:text=Answers%20in%20Genesis.-,Abstract,the%20absence%20of%20tail%20spikes[1].

**168. Ica or Nasca Burial Stones.**

These stones were made from 300 B.C. to 800 A.D. In addition to showing incredible ancient technology such as brain/heart surgeries and limb reattachment, these magnificent objects show that dinosaurs did indeed live alongside man. According to Dr. Hovind over 500 of these stones show dinosaurs. The real kicker is that some of these stones show circular patterns

---

1. https://answersresearchjournal.org/stegosaur-engravings-at-ta-prohm/#_853ae90f0351324bd73ea615e6487517__4c761f170e016836ff84498202b99827__853ae90f0351324bd73ea615e6487517_text_43ec3e5dee6e706af7766fffea512721_Answers_0bcef9c45bd8a48eda1b26eb0c61c869_20in_0bcef9c45bd8a48eda1b26eb0c61c869_20Genesis.-_c0cb5f0fcf239ab3d9c1fcd31fff1efc_Abstract_c0cb5f0fcf239ab3d9c1fcd31fff1efc_the_0bcef9c45bd8a48eda1b26eb0c61c869_20absence_0bcef9c45bd8a48eda1b26eb0c61c869_20of_0bcef9c45bd8a48eda1b26eb0c61c869_20tail_0bcef9c45bd8a48eda1b26eb0c61c869_20spikes

on the dinosaurs, this pattern was verified when fossilized Dino skin was found with that circle pattern.

Dinosaurs with Man The Evidence written in Stone? "Kent Hovind Creation Seminar 3, 48min"

### 169. Dinosaur pictographs and Petroglyphs.

Another fascinating find in regard to archaeology is the pictographs of dinosaurs made by Indians on the grand canyon, as well as petroglyphs made by the Anasazi in Blanding, Utah.

" Kent Hovind Creation Seminar 3, 44-45min"

### 170. Marco Polo & Job Posting

This famous guy reported that in 1271 A.D. the Chinese Emperor had dragons pull his chariots during parades. A position for "royal dragon feeder" was created by the Emperor in 1611. Dragon's blood was used for medicine and their eggs were highly prized.

" Kent Hovind Creation Seminar 3, 42/43min"

### 171. Alexander the great

Reportedly saw a hissing dragon/dragons in a cave in India in 326 B.C.

https://www.genesispark.com/exhibits/evidence/historical/dragons/

Creation seminar 3 by Kent Hovind 40min in.

### 172. 1946 Dictionary

A dictionary made in 1946 listed a dragon as a non-mythical creature (now rare).

"Kent Hovind Creation Seminar 3-Dinosaurs and the Bible, 32min"

### 173. Dragon's are mentioned in scripture and so is Behemoth.

Another evidence that dinosaurs lived with man is the occurrence of the word dragon/dragons made several times in the bible. Although many newer bibles omit some of the instances of these words, you can find them in the KJV. The Behemoth of Job 40 is undoubtedly a sauropod dinosaur. The word "dinosaur" wasn't made up until 1841 which is why it isn't found in the bible. The King James bible was made in 1611.

## 174. Length of Recorded History./ Problems with Egyptian king list (Manetho).

Another BIG problem against evolutionary theory is that the length of recorded history aligns almost perfectly with the account of creation given in the bible. If humans have been around for millions of years then how come we don't find artifacts dated to be in the tens of thousands of years range at the very least.

The most ancient books that we have are from the Hebrews and from the Egyptians. The dates assigned to Egyptian history is reckoned by king lists and going back as far as 3200 to 3600 B.C. However external as well as internal evidence suggest that these dated should be lowered. The issue we have with the dates of the first kings of Egypt is that they are based on a guy named Manetho. He was an Egyptian priests who lived in 250 B.C. Here is the problem with his dates. His writings are only preserved in some inaccurate quotes from other old writings.

The Evolution Handbook by Vance Ferrell pg 150-152

## 175. Diplodocus Sword.

This relic is now being held at the Arizona historical society. Discovered in the 1920's along with thirty other roman-style lead artifacts. This artifact appears to be genuine do to microscopic analysis of mineralization build up. As the heading implies, this is a sword with a diplodocus carved in it. If this relic was a hoax, what would be the motivation for making it? There would be valuable lead wasted and a considerable amount of burial work that would have had to be done.

https://www.genesispark.com/exhibits/evidence/historical/ancient/dinosaur/

### 176. Native American artifacts (pottery).

Now how on earth does Native American Artifacts refute evolutionary theory? Well the simple fact that they were found in the Grand canyon's uinkaret lava flows. These lava flows were dated to be 10,000 years old to 117 million years old, by Potassium Argon dating methods. Six tests were run, all failed to accurately date the age of the flows. Additionally, five Rubidium Strontium test dated the flows to over a BILLION years old! So they were wrong too. The Lead to Lead method gave a huge number as well for the age of the rocks ranging in the billions years old.

Dating Methods: CrEvo Rant 100

### 177. Ancient Peruvian vase depicts sauropod.

https://www.genesispark.com/exhibits/evidence/historical/ancient/dinosaur/

### 178. Bishop Bell's Tomb

A 15th century cathedral located in England contains a tomb that depicts brass carvings of dinosaurs.

" Kent Hovind Creation Seminar 3, 43min"

### 179. Ancient Technology.

Here is another brain-stumper for evolutionists. If man started out as an ape or ape-like creature and gradually go smarter over long periods of time, then why is it that ancient civilizations have amazing technologies (some of which we cannot duplicate today), whereas nowadays there is a great deal of people who are, let's say intellectually challenged (to put it nicely). There are a startling number of youth who do not know basic easy questions that virtually everybody over the age of 12 should know; questions that are on the level of easy as which 2 countries border the U.S? Here are just Ten examples of technology that shows the superiority of past cultures.

1. **Greek Fire**- Gel-like fire substance that could burn on water and very hard to put out.
2. **Lycurgus cup**- This ornate cup is amazing to look upon, it's glass

is diochroic, meaning it changes color when seen from different directions. When lit from behind it appears to be red in color but green when lit from the front. This is said to only be caused by nanotechnology (in this case the precise measurements of silver and gold nano-particles) nanotechnology supposedly wasn't created until the 1970's leaving archaeologists to assume it's creation was an accident.

3.  **Antikythera Mechanism**- A highly complex machine used to predict the movement of Celestial bodies with high accuracy. This artifact was found in a shipwreck off the coast of a Greek island. It was made around the 1st or 2nd century with clockwork that is so sophisticated the likes of which have only been made in the past few hundred years.

4.  **Sacsayhuaman**- Just outside of Cusco, Peru lies an Incan fort with walls made up of massive stacked stones weighing around 100 to 120 tonnes. Almost as impressive is that the surface of the stones appear to have a finish that is quite smooth and polished, this implies that the stones were melted and re-harden using immense heat. A natural fire could not do this, but are we to expect these people who lived in the 1400's to have made a kiln that could fit such massive stones.

5.  **Damascus Steel**- Has a reputation for being able to be formed into the sharpest edges as well as an incredible hardness, these blades were made during the middle ages in the middle east. The recipe has been lost to time. Modern day sword-smiths have attempted to recreate these swords but none claim to have 100% success.

6.  **Pyramids of Giza**- Comprised of millions of stones varying in weight of tens to hundreds of tonnes apiece, this magnificent structure has stumped the world for millennia. There have been many theories on how this structure could've been accomplished but so far no one has proven how they were made. There are various ramp theories, concrete theories, water theories, as well as theories including large complex machinery who designs have long since vanished. Although my source doesn't mention this let's not forget about the ancient aliens.;) Sarcastic wink emoji.

7.  **Archimedes screw**- Invented by Archimedes (a Greek engineer as well as a mathematician). Although the Egyptians probably used the

mechanism before his aha moment. This device is made by putting a helical screw inside a pipe/tube. The cool thing is, it is still being used to day in water and sewage treatment plants as well as in irrigation systems also.

8.  **Roman Concrete**- There are Roman structures that still stand today, (the most popular probably being the Roman Colosseum, thanks to Roman concrete. It's unique ingredients consisted of water, volcanic ash and lime.

9.  **The Baghdad Battery**- Thought to have originated around the 2nd or 3rd century, it consisted of a small clay jar with an iron rod and a copper cylinder divided by a layer of asphalt.

10. **The Phaistos Disc**- This disc is thought to originate around the 2nd century before Christ. It is covered in a bunch of pictures that seem to be quite indecipherable.

https://listverse.com/2021/10/15/10-ancient-technologies-we-cannot-recreate-today/

https://historycooperative.org/ancient-technology/

## 180. Vapor Canopy

The vapor Canopy hypothesis goes like this, long ago before Noah's flood there was a vapor canopy encircling the earth. This is possibly what Genesis 1:6-9 is referring to. This vapor canopy produced a more tropical climate globally than what we experience today. They probably didn't experience the drastic temperature changes that we experience today, I assume they didn't have cold snowy winters back then. The word winter isn't even mentioned in scripture until after the flood. We know that the present cannot be key to the past as evolutionist say because we find huge amounts of coal deposits in Antarctica (which is formed by vegetation and large trees that get buried). There is even evidence that enormous ferns and palm trees grew in the distant north and south. The fossil record also shows that the earth experienced a more uniform temperature in the past, as demonstrated by the similarity between arctic and non-arctic rock fossils. This may also explain why humans lived longer in the past, particularly in the first chapters of genesis (see also point #289). For more information on this topic, consider the resource below.

The Evolution Handbook by Vance Ferrell pg 647-652

**181. Language/Ancient Writing.**

Yet another fact that discredits the notion of molecules to man evolution is the conundrum of human speech and language. There is no such thing as a "primitive language". Our language is so more advanced and different than animal grunts and noises. There is no evidence that mankind's language has evolved from some primitive version. Animals do not name things with their speech, they do not make use of symbolic language, nor do they discuss things. Chimps may signal for things to get what they want but chimps do not have morals like us, nor do they describe things the way we do, nor is there a value system for them.

Mankind's language is amazing. No matter how far back we travel we see our language was already fully developed. Additionally, all our evidence points to the conclusion that writing did not occur until after 2500 B.C. What's more, is that as opposed to the idea that mankind gradually learned to speak and write from some sort of animal proto-language, studying earlier languages demonstrate that ancient man had a much, much more complicated writing system than modern man, and hence had more intelligence, not less.

For instance, ancient Greeks (300 A.D) would not use spaces between sentences or words, nor would they use paragraphs. Even more complicated than this 2500 years ago the Boustrophon writing style was used. In addition to no spacing, they wrote one line forward and the next line backward. In other words, they wrote from left to right then right to left. Ancient man was indeed smarter than us, no question. Sanskrit (Ancient language of India) was complicated as well, when they connected their words they changed the letter that joined them. For example our version of "how are you" in Greek is howareyou, but in Sanskrit is "howmrelou"

The Evolution Handbook by Vance Ferrell pg 582-86

## 182. Population Growth Statistics point to Biblical Narrative.

· 1999 A.D 6 Billion

· 1985 A.D 5 Billion

· 1810 A.D 1 Billion

· 2000 years ago 1/4 Billion

· Noah's day 8 people

If we have been here for three million years, we'd have **150** *people* <u>per sq. inch</u>.

" Kent Hovind Creation Seminar 1hr 5min"

## 183. Ancient Dinosaur Figurines?

Way back in 1944 a guy named Waldemar Julsrud was traveling by horseback around the base of El Toro mountain it was then that he spotted some ceramic pieces and carved stones sticking up out of the earth. He got off his horse and started diggin'. Assuming that there would be more of these things, he cut a bargain with a local farmer that for every piece he found he'd get 12 cents, a peso in Mexican money. During that time he found about 20,000 pieces. The artifacts that he found varied from mask, to people, to idols, to dinosaurs, yes, i said dinosaurs! Hundreds of dinosaur figurines were discovered. Each piece was unique. They were all hand-crafted.

Among some of the sculptures the T-Rex, Stegosaurus, Triceratops, and others could be identified. Now of coarse evolutionists will say that these figures are faked, a hoax, not genuine, etc. I also am open to the possibility they could be modern day forgeries. Still it is an interesting story that I thought I would mention is this book. While researching this piece of evidence of creation, I came across a video discussing the Acambaro figurines where I assume an evolutionists gave reasons why these figurines were a fake. He gave multiple reasons why he believed they weren't genuine. One good point he made (if it is true) is that the ceramics didn't show signs of being struck by a shovel or other excavation tool. We would expect to find some of them with scuffs, chips, etc if they were dug out of the ground. Also supposedly they didn't have dirt in their crevices or something along those lines.

On the other hand these Acambaro figurines are said to be impossible to fake because since the 1940's many more have been discovered by many people. Plus a man named Charles Hapgood is said to have proved their authenticity by digging underneath the concrete of a 1930's house and finding more of those figures. Although I am skeptical of this point of evidence in favor of Creation, nevertheless I have included it so that the reader may research this for themselves and come up with their own conclusions.

https://apologeticspress.org/dinosaur-figurines-in-mexico-4534/

### 184. The London Hammer.

The London Artifact discovered in June 1934, in London Texas. The discoverer's name was Max Hahn. He discovered this Cretaceous rock with wood protruding out of it. So he cracked it open with a hammer and chisel and lo and behold a metal hammer head! You could say this specimen throws a hammer in the gears of evolution.

https://www.creationevidence.org/displays/london_artifact.php

# Archaeology Confirms Scripture

### 185. Caiaphas' Ossuary.

An Ossuary is alot like a coffin. It holds dead men's bones. The specific ossuary that I'd like to mention here is that of Caiaphas, the high priests that interrogated Jesus. This bone box was discovered in 1990 during earth work on a hill south of Jerusalem.

https://dannythedigger.com/caiaphas-ossuary/

### 186. Chariot Wheels in the red sea?

I list this piece of evidence with caution. There is this man named Ron Wyatt, who has supposedly made nearly a hundred biblical archaeological discoveries within a decade. I list a website at the end of this point that questions the authenticity of such discoveries (as well as gives several reasons as to why one should be cautious of trusting Wyatt's research). Now why would I list a possibly fraudulent piece of evidence as support for the Christian worldview? While it's true that the supposed chariot wheels in the red sea may not be good evidence for the christian worldview because they may not be genuine; however they do serve as a reminder to the reader to not be gullible and believe everything that they hear.

Pro 14:15 The simple believeth every word: but the prudent *man* looketh well to his going.

1Th 5:21 Prove all things; hold fast that which is good.

Note: I am not saying that Wyatt's discoveries are fake or genuine, I will let my readers research and decide that for themselves. As my source says Archaeology is 10% excavation and 90% interpretation.

https://biblearchaeologyreport.com/2018/10/11/fake-news-in-biblical-archaeology/

## 187. The Jonah Sarcophagus.

Engraved in a stone sarcophagus is the story of the prophet Jonah, from the picture provided in the link below we see that Jonah was cast into the sea, swallowed by a sea monster, and then spit up, there is also the scene where Jonah was under the gourd. There are still other details inscribed on this object which you can check out for yourself from the website below.

https://www.christianiconography.info/sicily/sarcophagusJonah.html

## 188. Temple Inscription.

Back in the late 1960's an Archaeologist by the name of Benjamin Mazar found a stone block of the length of three feet that was among the ruins from the destruction of the second temple in 70 A.D. This stone block contained the inscription "To the place of trumpeting." This shows that it came from the location where the priests blew their trumpets.

https://biblearchaeologyreport.com/2019/01/19/top-ten-discoveries-in-biblical-archaeology-relating-to-the-new-testament/

## 189. Ketef Hinnom Scrolls.

Found in 1979 in a Judean burial cave, these two small metal and silver scrolls are said to be from the 7th century B.C. They contain the priestly benediction from Numbers 6, the Hebrew word Yahweh is also found inscribed during this discovery.

https://www.crossway.org/articles/10-crucial-archaeological-discoveries-related-to-the-bible/

## 190. Fulfilled prophecy.

One thing that makes the bible distinct from other religious text is the numerous and massive amounts of fulfilled prophesies it contains. Apart from divine inspiration there is no other plausible explanation for all these successful predictions of future events. For only God is outside of time and can say with accuracy what will certainly come to pass. My favorite type of prophecy in scripture are messianic prophesies. I have decided to list 70 of them below (in 14 sets of 5 verses apiece). These are also arranged in order from beginning to end of the bible (Old Testament). Additionally at the end of each verse, I will add the corresponding prophesy fulfilling verse reference. I've also included a link with hundreds more of such prophesies for your reading pleasure.

1. Gen 21:12 And God said unto Abraham, Let it not be grievous in thy sight because of the lad, and because of thy bondwoman; in all that Sarah hath said unto thee, hearken unto her voice; for in Isaac shall thy seed be called. *Hebrews 11:18*

Gen 22:8 And Abraham said, My son, God will provide himself a lamb for a burnt offering: so they went both of them together. *John 1:29*

Gen 22:18 And in thy seed shall all the nations of the earth be blessed; because thou hast obeyed my voice. *Galatians 3:16*

Num 24:17 I shall see him, but not now: I shall behold him, but not nigh: there shall come a Star out of Jacob, and a Sceptre shall rise out of Israel, and shall smite the corners of Moab, and destroy all the children of Sheth. Galatians 4:4

Deu 32:21 They have moved me to jealousy with *that which is* not God; they have provoked me to anger with their vanities: and I will move them to jealousy with *those which are* not a people; I will provoke them to anger with a foolish nation. *Romans 11:11*

2. 2Sa 7:16 And thine house and thy kingdom shall be established for ever before thee: thy throne shall be established for ever. Revelations 22:16

Job 19:26 And *though* after my skin *worms* destroy this *body,* yet in my flesh shall I see God: *John 5:24-29*

Psa 2:7 I will declare the decree: the LORD hath said unto me, Thou *art* my Son; this day have I begotten thee. *Matthew 3:17*

Psa 8:2 Out of the mouth of babes and sucklings hast thou ordained strength because of thine enemies, that thou mightest still the enemy and the avenger. *Matthew 21:16*

Psa 16:10 For thou wilt not leave my soul in hell; neither wilt thou suffer thine Holy One to see corruption. *Acts 2:31*

3. Psa 22:1 To the chief Musician upon Aijeleth Shahar, A Psalm of David. My God, my God, why hast thou forsaken me? *why art thou so* far from helping me, *and from* the words of my roaring? *Mark 15:34*

Psa 22:8 He trusted on the LORD *that* he would deliver him: let him deliver him, seeing he delighted in him. *Matthew 27:43*

Psa 22:14 I am poured out like water, and all my bones are out of joint: my heart is like wax; it is melted in the midst of my bowels. *John 19:34*

Psa 22:15 My strength is dried up like a potsherd; and my tongue cleaveth to my jaws; and thou hast brought me into the dust of death. *John 19:28*

Psa 22:16 For dogs have compassed me: the assembly of the wicked have inclosed me: they pierced my hands and my feet. John 20:27

4. Psa 22:17 I may tell all my bones: they look *and* stare upon me.

Psa 22:18 They part my garments among them, and cast lots upon my vesture. *Luke 23:34-35*

Psa 31:5 Into thine hand I commit my spirit: thou hast redeemed me, O LORD God of truth. *Luke 23:46*

Psa 34:20 He keepeth all his bones: not one of them is broken. John 19:31-36

Psa 55:12-13 For *it was* not an enemy *that* reproached me; then I could have borne *it:* neither *was it* he that hated me *that* did magnify *himself* against me; then I would have hid myself from him: But *it was* thou, a man mine equal, my guide, and mine acquaintance. *John 13:18*

5.Psa 68:18 Thou hast ascended on high, thou hast led captivity captive: thou hast received gifts for men; yea, *for* the rebellious also, that the LORD God might dwell *among them. Luke 24:51*

Psa 68:18 Thou hast ascended on high, thou hast led captivity captive: thou hast received gifts for men; yea, *for* the rebellious also, that the LORD God might dwell *among them. Ephesians 4:7-16*

Psa 69:9 For the zeal of thine house hath eaten me up; and the reproaches of them that reproached thee are fallen upon me. *John 2:17*

Psa 69:20 Reproach hath broken my heart; and I am full of heaviness: and I looked *for some* to take pity, but *there was* none; and for comforters, but I found none. *Matthew 26:38*

Psa 69:21 They gave me also gall for my meat; and in my thirst they gave me vinegar to drink. Mat 27:34

6. Psa 72:10 The kings of Tarshish and of the isles shall bring presents: the kings of Sheba and Seba shall offer gifts. *Matthew 2:1-11*

Psa 78:2 I will open my mouth in a parable: I will utter dark sayings of old: *Matthew 13:34-35*

Psa 109:4 For my love they are my adversaries: but I *give myself unto* prayer. *Luke 23:34*

Psa 118:22-23 The stone *which* the builders refused is become the head *stone* of the corner. This is the LORD'S doing; it *is* marvellous in our eyes. *Matthew 21:42-43*

Pro 1:23 Turn you at my reproof: behold, I will pour out my spirit unto you, I will make known my words unto you. *John 16:7*

7. Isa 6:9 And he said, Go, and tell this people, Hear ye indeed, but understand not; and see ye indeed, but perceive not. *Matthew 13:13-15*

Isa 7:14 Therefore the Lord himself shall give you a sign; Behold, a virgin shall conceive, and bear a son, and shall call his name Immanuel. *Luke 1:35*

Isa 9:6 For unto us a child is born, unto us a son is given: and the government shall be upon his shoulder: and his name shall be called Wonderful, Counsellor, The mighty God, The everlasting Father, The Prince of Peace. *Matthew 13:54*

Isa 40:3 The voice of him that crieth in the wilderness, Prepare ye the way of the LORD, make straight in the desert a highway for our God. *John 1:23*

Isa 42:7 To open the blind eyes, to bring out the prisoners from the prison, *and* them that sit in darkness out of the prison house. *John 9:25-38*

8. Isa 48:12 Hearken unto me, O Jacob and Israel, my called; I *am* he; I *am* the first, I also *am* the last. *Revelation 1:8*

Isa 49:6 And he said, It is a light thing that thou shouldest be my servant to raise up the tribes of Jacob, and to restore the preserved of Israel: I will also give thee for a light to the Gentiles, that thou mayest be my salvation unto the end of the earth. *Acts 13:47*

Isa 50:6 I gave my back to the smiters, and my cheeks to them that plucked off the hair: I hid not my face from shame and spitting. *Matthew 27:6*

Isa 53:3 He is despised and rejected of men; a man of sorrows, and acquainted with grief: and we hid as it were *our* faces from him; he was despised, and we esteemed him not. *Matthew 27:21-23*

Isa 53:4 Surely he hath borne our griefs, and carried our sorrows: yet we did esteem him stricken, smitten of God, and afflicted. *Luke 6:17-19*

9. Isa 53:5 But he *was* wounded for our transgressions, *he was* bruised for our iniquities: the chastisement of our peace *was* upon him; and with his stripes we are healed. *Colossians 1:20*

Isa 53:6 All we like sheep have gone astray; we have turned every one to his own way; and the LORD hath laid on him the iniquity of us all. *Galatians 1:4*

Isa 53:7 He was oppressed, and he was afflicted, yet he opened not his mouth: he is brought as a lamb to the slaughter, and as a sheep before her shearers is dumb, so he openeth not his mouth. *Matthew 27:12-14*

Isa 53:9 And he made his grave with the wicked, and with the rich in his death; because he had done no violence, neither *was any* deceit in his mouth. *Matthew 27:57*

Isa 53:10 Yet it pleased the LORD to bruise him; he hath put *him* to grief: when thou shalt make his soul an offering for sin, he shall see *his* seed, he shall prolong *his* days, and the pleasure of the LORD shall prosper in his hand. *Matthew 20:28*

10. Isa 53:11 He shall see of the travail of his soul, *and* shall be satisfied: by his knowledge shall my righteous servant justify many; for he shall bear their iniquities. *Romans 5:8-9*

Isa 53:12 Therefore will I divide him *a portion* with the great, and he shall divide the spoil with the strong; because he hath poured out his soul unto death: and he was numbered with the transgressors; and he bare the sin of many, and made intercession for the transgressors. *Luke 23:34*

Isa 61:1 The Spirit of the Lord GOD *is* upon me; because the LORD hath anointed me to preach good tidings unto the meek; he hath sent me to bind up the brokenhearted, to proclaim liberty to the captives, and the opening of the prison to *them that are* bound; *John 8:31-32*

Isa 63:2 Wherefore *art thou* red in thine apparel, and thy garments like him that treadeth in the winefat? *Revelation 19:13*

Jer 30:9 But they shall serve the LORD their God, and David their king, whom I will raise up unto them. *John 18:37*

11. Jer 31:15 Thus saith the LORD; A voice was heard in Ramah, lamentation, *and* bitter weeping; Rahel weeping for her children refused to be comforted for her children, because they *were* not. *Matthew 2:16-18*

Jer 31:31 Behold, the days come, saith the LORD, that I will make a new covenant with the house of Israel, and with the house of Judah: *Matthew 26:28*

Dan 7:13 I saw in the night visions, and, behold, *one* like the Son of man came with the clouds of heaven, and came to the Ancient of days, and they brought him near before him. *Mark 14:62*

Dan 7:14 And there was given him dominion, and glory, and a kingdom, that all people, nations, and languages, should serve him: his dominion *is* an everlasting dominion, which shall not pass away, and his kingdom *that* which shall not be destroyed. *Luke 1:33*

Dan 9:24 Seventy weeks are determined upon thy people and upon thy holy city, to finish the transgression, and to make an end of sins, and to make reconciliation for iniquity, and to bring in everlasting righteousness, and to seal up the vision and prophecy, and to anoint the most Holy. *Galatians 1:3-5*

12. Dan 9:25 Know therefore and understand, *that* from the going forth of the commandment to restore and to build Jerusalem unto the Messiah the Prince *shall be* seven weeks, and threescore and two weeks: the street shall be built again, and the wall, even in troublous times. *John 12:12-13*

Dan 9:26 And after threescore and two weeks shall Messiah be cut off, but not for himself: and the people of the prince that shall come shall destroy the city and the sanctuary; and the end thereof *shall be* with a flood, and unto the end of the war desolations are determined. *Hebrews 2:9*

Hos 11:1 When Israel *was* a child, then I loved him, and called my son out of Egypt. *Matthew 2:14*

Hos 13:14 I will ransom them from the power of the grave; I will redeem them from death: O death, I will be thy plagues; O grave, I will be thy destruction: repentance shall be hid from mine eyes. *1 Corinthians 15:55*

Joe 2:32 And it shall come to pass, *that* whosoever shall call on the name of the LORD shall be delivered: for in mount Zion and in Jerusalem shall be deliverance, as the LORD hath said, and in the remnant whom the LORD shall call. *Romans 10:13*

Amo 8:9 And it shall come to pass in that day, saith the Lord GOD, that I will cause the sun to go down at noon, and I will darken the earth in the clear day: *Acts 2:20*

13. Jon 1:17 Now the LORD had prepared a great fish to swallow up Jonah. And Jonah was in the belly of the fish three days and three nights. *Luke 24:7*

Mic 5:2 But thou, Bethlehem Ephratah, *though* thou be little among the thousands of Judah, *yet* out of thee shall he come forth unto me *that is* to be ruler in Israel; whose goings forth *have been* from of old, from everlasting. *Matthew 2:1*

Nah 1:7 The LORD *is* good, a strong hold in the day of trouble; and he knoweth them that trust in him. *2 Timothy 2:19*

Hag 2:23 In that day, saith the LORD of hosts, will I take thee, O Zerubbabel, my servant, the son of Shealtiel, saith the LORD, and will make thee as a signet: for I have chosen thee, saith the LORD of hosts. *Matthew 1:13*

Zec 9:9 Rejoice greatly, O daughter of Zion; shout, O daughter of Jerusalem: behold, thy King cometh unto thee: he *is* just, and having salvation; lowly, and riding upon an ass, and upon a colt the foal of an ass. Matthews 21:6-9

14. Zec 11:12 And I said unto them, If ye think good, give *me* my price; and if not, forbear. So they weighed for my price thirty *pieces* of silver. *Matthew 26:14-15*

Zec 11:13 And the LORD said unto me, Cast it unto the potter: a goodly price that I was prised at of them. And I took the thirty *pieces* of silver, and cast them to the potter in the house of the LORD. *Matthew 27:3-5*

Zec 12:10 And I will pour upon the house of David, and upon the inhabitants of Jerusalem, the spirit of grace and of supplications: and they shall look upon me whom they have pierced, and they shall mourn for him, as one mourneth for *his* only *son,* and shall be in bitterness for him, as one that is in bitterness for *his* firstborn. *John 19:34-37*

Zec 13:7 Awake, O sword, against my shepherd, and against the man *that is* my fellow, saith the LORD of hosts: smite the shepherd, and the sheep shall be scattered: and I will turn mine hand upon the little ones. *Matthew 26:31-56*

Mal 4:5 Behold, I will send you Elijah the prophet before the coming of the great and dreadful day of the LORD: Matthew 3:1-2

https://www.hopefaithprayer.com/scriptures/old-testament-prophecies-jesus/

## 191. Flood Legends.

There are more that 270 stories/historical accounts of the flood worldwide. Some of these stories are remarkable similar to the Genesis account. The Hawaiian account states that the world became a wicked place after the death of the first man, then the only good man (Nu-u) left built a canoe filled it with animals and only he and his family were saved. According to the Chinese classic "Hiking", a man Fuhi survived a massive flood with his family (wife, 3 sons and their wives). They were the only people on earth and they repopulated the planet.

" Kent Hovind Creation Seminar 3, 17min"

## 192. The Empty Tomb.

Atheists and skeptics have no good explanation for the empty tomb.

The fact is if the Jewish leader or Romans took the body, they would have used it to dispel the rumors of the resurrection. The disciples would not have stole the body because they were taught it was wrong to lie and steal (the ten commandments); plus they would not have endured the persecution and deaths that they faced for a lie. Even if they wanted to steal the body, they would have to deal with the guards and the giant stone. If Jesus' didn't rise from the dead, we wouldn't be able to explain his appearance for forty days after he was risen as well as the fact that he appeared to 500+ witnesses at once (1st Corinthians 15:6).

## 193. Ancient Chinese Characters depict Genesis.

In turns out that some of the details of Genesis can be found in the ancient Chinese language. For this section I will just give five examples; although it shouldn't be hard for my readers to find more if they are interested.

1. The Chinese character for trees + the character for woman = **desire/covet**.

2. The Chinese character for dust + breath + two people + enclosure = **garden**.

3. Eight + united + earth + total + water = **flood.**

4. Vessel + eight + mouth = **boat**.

5. The words tongue + walking = **rebellion/confusion**. The tower of babel.

The Evolution Handbook by Vance Ferrell pg 642

## 194. Elephantine Papyrus

Sanballat and Tobiah are mentioned several times in these papri. Sanballat and Tobiah were two enemies of Israel and are found in the book of Nehemiah.

https://aish.com/48969466/

## 195. Babylonian Chronicles

...verify that Nebucadnessar did indeed install the last Judean king, Zedekiah.

https://aish.com/48969466/

## 196. Testimony of Scripture

The old earth view has no support in scripture whatsoever. I could be mistaken but I believe there are only a handful of passages old earth creationist can pull from and twists to try and cram millions of years into the bible. I will get to why the bible teaches a young earth later in this book but for know I will list the two passages that I'm aware of that old earth creationists twists in order to teach the bible promotes millions of years. These theories cannot work because that would put death before sin, which would in turn be blaming God for death. This would be an attack on the character of God.

### <u>Day age theorists use:</u>

2Pe 3:8 But, beloved, be not ignorant of this one thing, that one day *is* with the Lord as a thousand years, and a thousand years as one day.

Day-agers like to use the first part of this verse in order to support their theory that each day of creation week was an indefinite amount of time, but the second half of the verse refutes their claim I think. Plus how did plant survive so long without the sun?

**<u>Gap theorists use:</u>**

Jer 4:23 I beheld the earth, and, lo, *it was* without form, and void; and the heavens, and they *had* no light.

Jer 4:24 I beheld the mountains, and, lo, they trembled, and all the hills moved lightly.

Jer 4:25 I beheld, and, lo, *there was* no man, and all the birds of the heavens were fled.

Jer 4:26 I beheld, and, lo, the fruitful place *was* a wilderness, and all the cities thereof were broken down at the presence of the LORD, *and* by his fierce anger.

Jer 4:27 For thus hath the LORD said, The whole land shall be desolate; yet will I not make a full end.

Gap theorists will then point the word <u>*void*</u> and say that in Hebrew it means ruins or something along those lines. Sure that is one definition according to Strong's Hebrew (an indistinguishable ruin). But it also can mean empty or emptiness, or void. So why pick the definition they want if that's all the evidence they got from scripture for their theory.

## 197. Tel Dan Inscription.

This was uncovered by excavationists in 1993. Inscribed are the words BYTDWD. It is believed this word means "house of David", this artifact is said to come from the 9th century B.C.

https://www.crossway.org/articles/10-crucial-archaeological-discoveries-related-to-the-bible/

## 198. Pliney the Younger.

Pliney references the Lord's supper as well as states that the early Christians were very ethical and worshiped Jesus as God.

https://www.gotquestions.org/did-Jesus-exist.html

## 199. Flavius Josephus.

A very well known Jewish historian. Some things Josephus wrote about or mentioned...

· James the brother of Jesus

· Jesus did miracles (wrought surprising feats)

· Jesus' Resurrection

· Jesus' conduct was good and he was virtuous.

https://www.gotquestions.org/did-Jesus-exist.html

## 200. The Talmud.

Mentions Jesus. Although not in a positive light. It accuses Jesus of sorcery and encouraging apostasy, however it does say that Jesus was crucified.

https://www.gotquestions.org/did-Jesus-exist.html

## 201. Tacitus.

Tacitus was a first century Roman who mentioned that Christians were distressed under Pontius Pilate when Tiberius reigned. Tacitus is considered one of the more reliable historians of long ago.

https://www.gotquestions.org/did-Jesus-exist.html

## 202. Polycarp.

Polycarp was a contemporary of Ignatius, a disciple of the Apostle John, and the one who taught Ireneaus. Polycarp was born in the 2nd half of the first century A.D, and continued until around the middle of the second century. He is considered to be one of the Apostolic Fathers (who were the leaders of the church right after the Apostles). He was to be burnt at the stake, but it is said the flames avoided him, so that he was killed by spear, his blood is said to have extinguished the flames.

https://www.gotquestions.org/Polycarp.html

## 203. Moabite Stone.

A missionary found this stone in 1868, the stone was broken in pieces to make more money, but thankfully a copy was made before that. The stone dates way back to the ninth century B.C. The stone's topic is about King Mesha's victories over Israel in capturing cities under their control. The bible (2 Kings 3) focuses on Israel's successful counter attacks against Moab. So although the perspectives are different, this stone verifies biblical events.

https://www.crossway.org/articles/10-crucial-archaeological-discoveries-related-to-the-bible/

## 204. The Bible No contradiction in scripture.

One of the greatest things about the bible is that it is true; and as such their are no contradictions found therein. Now you can find websites that claim there are contradictions in the bible but these conclusions are based on faulty logic and unwarranted assumptions.

The bible was written by 40 or more different authors over a period of 1600 years, and yet the books of bible is so perfectly united in core truths. What's amazing is that although there are so many supposed contradictions found on the internet, the vast majority can be solved using logic and reason. Very few supposed contradictions can be said to be a mystery. The point is the answers are out there if you want to know the truth.

If you want to know how to solve difficult supposed bible contradictions I recommend "Keeping Faith in an age of Reason" by Jason Lisle. https://defendinginerrancy.com[2] is another resource that you can check out, but it is free.

## 205. Rahab's house.

Back in the early 1900's, Germans excavated Jericho and found that a portion of the wall has still remained till that present day. Additionally there were houses that were constructed on a rampart, that located against this wall. The walls of these houses were not thick but was made up of only one brick thick which

---

2. https://defendinginerrancy.com/

indicated that this area of the city was not rich. Lending to the probability that this was indeed Rahabs house.

https://biblearchaeologyreport.com/2019/05/25/biblical-sites-three-discoveries-at-jericho/

## 206. Azariah Bulla.

A bulla is a seal that has been attracted to an official document; one of such seals was found in Jerusalem it bears the name **Azariah son of Hilkiah** confirming the existence of Ezra's grandfather (1 Chronicles 5 and Ezra 7).

https://aish.com/48969466/

## 207. Ancient Plaster

Fragments contain "Warnings given by Balaam, the son of Beor. A seer of the gods."

https://www.biblehistory.net/balaam.html

## 208. Peter's House

Undoubtedly, one of the most famous of Jesus' disciples was the apostle Peter. Is there any evidence of his existence outside the bible though? As a matter of fact, yes. The remains of Peter's house can be found in Capernaum. We found out this was where Peter lived from reading Matthew 8. How do we know that this was Peter's house. There's his name on the walls and we also have discovered that his house was eventually turned into a church. From the link I have provided, I suppose this site contained large vessels used to cater to crowds. A giant boat-shaped church building has been built over the site.

https://storiesofourboys.com/2023/05/08/10-exciting-archaeology-finds-that-support-the-bible-2/

## 209. Jezebel.

Perhaps the most wicked women of all of the bible, Jezebel's name was found on a Phoenician royal seal which has been dated to the time of her Husband Ahab.

https://aish.com/48969466/

## 210. King David's Palace.

One of the most wrote about person in the bible is King David. King David lived around the 10th century B.C. So is there any archaeological evidence for the existence of this famous king? Excavator Eilat Mazer and author Nadav Na'aman believe so. South of the Temple Mount in the city of David lies two notable stone monuments. One is a massive stone structure, the other a stepped stone structure. They are believed to be the Palace of the famous king and the millo. This conclusion is the result of using both archaeological as well as textual evidence. This impressive building was made up of ashlar blocks. Na'aman believes this is indeed Davad's palace due to the scale, the date, as well as the location of the structure. Also found near the stone remains were two bullae of Judahite officials. One bullae has the inscription of Gedaliah son of Pashhur, the second bullae contains the name Jehucal son of Shelemiah son of Shobai. These were king Zedikaiah's men according to scripture (Jeremiah 37:3, 38:1).

see also: 2 Samuel 5:11; 2 Samuel 11; 2 Samuel 16:22,Nehemiah 12:37

https://www.biblicalarchaeology.org/daily/biblical-sites-places/jerusalem/king-davids-palace-and-the-millo/

## 211. Hezekiah's tunnel.

Found in 1867 by a man named Charles Warren. This water shaft allowed the inhabitants of Jerusalem to drink from the Gihon spring during times of seige, it was needed because the spring is located outside city walls.

https://www.crossway.org/articles/10-crucial-archaeological-discoveries-related-to-the-bible/

## 212. Annals of Esarhaddon and the annals of Assurbanipal.

Attest to the Nefarious Judean king Manasseh, the son of Hezekiah.

https://aish.com/48969466/

### 213. Pool of Siloam.

This is yet another evidence for the veracity of scripture. It was discovered around 20 years ago when a water pipe busted. It's location: Lower region of the city of David, on a main road that went up to the temple's south gate. It was as big as a couple Olympic size swimming pools. It's water came from the Gihon spring. Jesus healed a blind man using this pool.

Joh 9:6 When he had thus spoken, he spat on the ground, and made clay of the spittle, and he anointed the eyes of the blind man with the clay,

Joh 9:7 And said unto him, Go, wash in the pool of Siloam, (which is by interpretation, Sent.) He went his way therefore, and washed, and came seeing.

https://www.holylandsite.com/pool-of-siloam

### 214. Ark Dimensions

This is really incredible! How can we possibly explain how Noah knew the best dimensions to build a super seaworthy vessel would be?

The dimensions were 300 cubits long, 50 wide, and 30 high. In feet that would be 450 feet long, by 75 feet wide, and 45 feet high. Don't think that these were good measurements? Then explain why almost the exact same measurements were used to build the Great Britain by Isambard K. Brunnel in 1844, and again others used it during ww2 to build the S.S Jeremiah O' Brien.

https://apologeticspress.org/noahs-arka-flawless-floater-562/

### 215. Dead Sea Scrolls.

Discovered in a cave by shepherds in 1947, these writings (mostly on leather but some on parchment), contain parts nearly every OT book. A complete scroll of Isaiah was also discovered. Over 800 fragmentary documents were discovered (however not all were of the scriptures). The dead sea scrolls were dated to be from the 1st to the 3rd century B.C. These scrolls were touted to be the greatest 20th century find in archaeology. The greatest thing about these scrolls is that they show how similar the OT writings back then are to the OT texts we have today.

https://www.crossway.org/articles/10-crucial-archaeological-discoveries-related-to-the-bible/

# Ch. 5 The Heavens Declare His Glory: Evidence from Astronomy

## How Cosmic Evolution Defies Physics

### 216. Conservation of angular momentum.

Some Evolutionist may tell you that the universe came from a really small, really hot spinning dot. Well then if that were true we should expect the planets and moons to be spinning in the same direction because according to the Conservation of angular momentum the things which break off a spinning object would be spinning in the same direction until they faced resistance; but we have backwards spinning planets (Venus & Uranus), moons, and there are even backwards spinning galaxies.

The Evolution Handbook by Vance Ferrell pg 84,105, " Kent Hovind Creation Seminar 1 31-34min"

### 217. You can't get from dust clumps to planets.

According to the Evolutionist model of the origin of our solar system, the sun form first, then the swirling dust that orbited the sun clumped together to form dust clumps, then those clumps smooshed together to form small rocks, then the small rocks clumped together to form bigger rocks, then asteroids (planetesimals), then finally the planetesimals clumped together to form our planets.

The biggest problem with this theory is that it's not possible. Once you get to dust clumps or small pebbles the dust clump/pebbles would hit each other too fast and break up. It is only when they are small asteroids could the gravity cause them to stick together.

"Our Created solar system" Narrated by Spike Psarris [Youtube]

### 218. Boyle's Gas law.

According to evolutionary theory gas in outer space clumped together to form stars. Here is the dilemma. Gas in space is even less compact than on earth. If gas doesn't clump together here on earth. How can it do it in space? Gas spread out, it doesn't clump together; that's just science.

The Evolution Handbook by Vance Ferrell pg 74-75

# Clearly Designed

### 219. Flatness Problem.

Despite the fact that there could have been an extremely wide range of possibilities, the expansion rate of the universe and gravity are extremely finely balanced. At the moment of the supposed " big bang" those two forces would have had to been very finely in balance in order for us to have the universe the way that we witness at our present time.

"Big Problems with the Big Bang Theory" by Jason Lisle [Youtube]

### 220. Horizon Problem.

Very often we have evolutionist say to YEC (young earth creationists) that the mere fact that we can see starlight shows the universe is millions of years old because that is how long the light would take to get to us. While there are theories to explain distant starlight from a creationists view, one being that God could have made the light arriving to earth from the very beginning. After all God made the garden, animals and man mature from the beginning. The Evolutionists cannot explain however why the cosmic microwave background (CMB) has such an even temperature throughout the visible universe. Light should not have had enough time to exchange heat throughout the universe even given the supposed 16+ billion year existence of the universe.

"Big Problems with the Big Bang Theory" by Jason Lisle [Youtube]

### 221. Mono-pole Problem.

The absence of magnetic mono-poles pose a significant problem for the big bang theory because big bang temperatures (super hot) should have produces them but they (magnetic mono-poles) have not been found.

"Big Problems with the Big Bang Theory" by Jason Lisle [Youtube]

### 222. Singularity problem.

A singularity is the description of the beginning of the universe right before the supposed big bang, which would have been infinite temperature & density but no space.

Problems with a singularity

- Doesn't explain the cause of the universe

- the mere existence of a singularity (infinity) usually means you goofed up on your physics problem

- Physics breaks down at the singularity level

"Big Problems with the Big Bang Theory" by Jason Lisle [Youtube]

## 223.Singing Stars.

Asteroseismology is study of star quakes, these star quakes generate sound waves in the star. Unlike earthquakes, star quakes happen all the time in stars. Although we cannot hear these sound waves through space we can however use advanced telescopes in order to discern what the sounds are like based on the differing brightness of a star. Sometimes the wavelengths of these stellar oscillations (which are grafted as functions of time) are in the period of years, so in order to make them audible to humans we have to shift the wavelengths, this process is known as sonification.

This information was derived from a youtube video titled: Asteroseismology: The Sound of the Stars by Grace Davis.

1Co 15:41 *There is* one glory of the sun, and another glory of the moon, and another glory of the stars: for *one* star differeth from *another* star in glory.

P.S. don't wish upon a star, that's idolatry.

While we are on the topics of stars I would just like to state that distant starlight is not a problem for the young earth model of creation. There are multiple explanations as to how we can see starlight supposedly millions of light years away. God could have made the universe fully mature like he did the rest of creation (Adam and Eve were made adults) and Christians usually suppose the animals in the garden were made mature as well. Another explanation is the fact that God stretched out the heavens, which may have included the starlight as well.

Isa 44:24 Thus saith the LORD, thy redeemer, and he that formed thee from the womb, I *am* the LORD that maketh all *things;* that stretcheth forth the heavens alone; that spreadeth abroad the earth by myself;

Zec 12:1 The burden of the word of the LORD for Israel, saith the LORD, which stretcheth forth the heavens, and layeth the foundation of the earth, and formeth the spirit of man within him.

### 224. Antimatter.

Antimatter poses two massive problems for the theory of evolution. First, if the big bang really occurred it should have produced an equal amount of antimatter, but little to none exists. Secondly if the big bang really did happen and did produce antimatter, the antimatter and the regular matter would have destroyed one another. That's what happens in the lab.

The Evolution Handbook by Vance Ferrell pg 72-73

### 225. Venus.

*Why is Venus so different from Earth?* It's covered in lava flows and rains sulfuric acid. Temperatures get up to 900 degrees and is covered with clouds which are mostly Carbon Dioxide. Venus also has 90 atmospheres of pressure. Venus has no magnetic field or tectonic plates, and no moon.

According to Evolution Venus;...

- Was formed from the same materials as earth

- At the same time

- by the same Natural Processes

- And about the same place

Venus supposedly formed out of the same kinds of materials and around the same time as earth, why is it so different? Venus' surface also appears quite young. Evolutionist claim the entire planet was resurfaced around 500 million years ago. Although this is going by their flawed cratering model.

"Our Created solar system" Narrated by Spike Psarris [Youtube]

## 226. The existence of our Moon.

What is it about the moon that disproves evolution? Evolutionary Scientists have to good explanation of how our moon got where it was. There are at least three evolutionary theories to explain the moon's existence but they all have huge problems with them. Let's take a look at all three, the problems are listed as bullet points.

1. The Fission Theory-states when the earth was young it was spinning very very fast, then a part of the earth flew off and became our moon.

· Earth's rocks are different than the moon's.

· The rotation of the Earth would've been ten times faster than the current speed, this speed would have increased the temperature of the earth by a thousand degrees C°. We know the fission theory is incorrect because this temperature increase would have evaporated our oceans.

2. The Nebula Theory- Both the earth and moon was created from gas and dust.

· Earth's rocks are different than the moon's.

· As stated earthier, we cannot get from dust clumps to planets, or moons for that matter.

3. The Capture Theory- the moon was traveling from somewhere else and it got stuck in earth's orbit.

· The moon would have speed up by earth's gravity and would have hit the earth or passed on by.

"Our Created solar system" Narrated by Spike Psarris [Youtube]

### 227. The Eclipse.

It just so happens that the sun is 400x bigger than the moon, but is also 400x father away! This creates a perfect eclipse, and it just so happens to be here where we can observe it.

"Our Created solar system" Narrated by Spike Psarris [Youtube]

### 228. Miranda.

This very small moon of Uranus, is only about 300 miles in diameter. It's surface is so bizarre, and resembles a "patchwork quilt" as more source states. You guy's really owe it to yourselves to at least check out a picture of this strange moon, so many different terrains, it's incredible. Evolutionists propose that Miranda was hit by five asteroids over the course of it's evolution. *cough* bologna *cough*.

"Our Created solar system 1hr 20min" Narrated by Spike Psarris [Youtube]

### 229. Dancing Moons.

Saturn's moons (Janus and Epimetheus) switch orbits every four years

"Our Created solar system 1hr 20min" Narrated by Spike Psarris [Youtube]

## *Youthful Appearance Indicates Recently Created*

### 230. Early faint Sun Paradox

According to evolutionary theory the sun has used up almost half it's energy reserves in the 4.6 billion years of it's supposed existence. This would mean that half of the hydrogen in it's core has been used up and converted into the element helium. This would in turn effect the sun's luminosity, it should be 40% brighter then it was at it's creation. According to evolution the temperature on earth when life first appeared must have been similar to today's temperature in order for life to evolve, and yet this means that it would be too hot for life today. Some may suggest it was colder in the past and we are gradually getting hotter, geologist claim the rock record says Earth's temperature has not varied wildly in the past.

https://www.icr.org/article/429/

## 231. Mercury.

Mercury has a Magnetic Field, this creates a huge problem for evolutionist. This is because magnetic fields shouldn't last billions of years. The only explanation for why Mercury has one would be a dynamo in Mercuries core. However the problem for evolutionist is, Mercury is thought to have a solid core which would prevent a dynamo.

"Our Created solar system" Narrated by Spike Psarris [Youtube]

## 232. Earth's Rotation.

The earth's rotation is slowing down. Currently it is going about 1000mph. If the earth was truly billion's of years old, like many scientists say, then the earth should have stopped spinning on it's axis.

"The Evolution Handbook by Vance Ferrell pg. 138-139"

## 233. Moon Dust

If billions of years old there should have been 20 to 60 mile layer of dust on surface, turns out only 2 or 3 inches. This is consistent with the biblical age of 6 to 8 thousand years.

The Evolution Handbook by Vance Ferrell pgs 131,132

## 234. Receding moon.

The moon gets 1.5 inches farther away from the earth per year. Meaning in the past it was much closer. Just twenty or thirty thousand years ago the moon would have crashed into the earth.

The Evolution Handbook by Vance Ferrell pg 134, "Our Created solar system"

## 235. The Moon's Ghost Craters.

The moon has crater's and then it has what's known has ghost craters. Ghost craters are formed when large craters (by huge impacts) are soon followed up by additional smaller craters crashing into the larger ones, and then the small craters get partially filled up with lava, leaving only the barely visible top of them exposed for observation. We wouldn't get this effect it it took thousands or millions of years for all these craters to occur. This calls into question all the assigned ages of other moons in our solar system that are based on the long age cratering modal.

https://creation.com/age-of-the-earth

## 236. Jupiter.

This large planet emits almost twice as much energy that it gets from the sun. If Billions of years old, why hasn't it cooled off yet?

https://www.icr.org/article/7834

## 237. Europa's surface, too few craters.

Almost perfectly smooth! The moon is en-wrapped in an ice shell. It turns out that that one object crashing into a planet can produce a lot of debris and form many small or medium craters. It turns out that up to a whopping 95% of craters formed in this fashion. Less impacts means less time.

"Our Created solar system" 1hr 4min Narrated by Spike Psarris [Youtube]

## 238. Io's Heat.

Io has 400+ volcanoes. These volcanoes have been seen blasting material 180 miles into outer space. Io also puts out an astounding amount of heat, which is twice as much as earth's. This is partially due to tidal flexing which is caused by the gravitation pull of Jupiter and it's other moons. If Io is was recently created, then it could still be cooling down from it's initial formation. Another thing about Io's volcanoes is that they are super active, they are erupting very frequently. It turns out that Io would have recycled itself through it's own volcanoes well over two dozen times, if it was 4.5 billion years. Io's lava is over a thousand degrees hotter than earth's (Io's lava is 3000f). On top of that the fact that this material is high density proves Io isn't even millions of years old. As the high density lava material would have sunk down to the inside if millions of years old.

"Our Created solar system 1hr 6min" Narrated by Spike Psarris [Youtube]

**239.Saturn's Rings.**

According to planetary scientist Jeff Cuzzi from the NASA Ames Research Center, Saturn's rings look new. They look shiny, which is not to be expected if millions of years old, they should have collected space dust and turned to a charcoal color under the old universe model. What's more Saturn's moons (the one's among the rings), should have darted elsewhere.

https://www.icr.org/article/planetary-quandaries-solved-saturn

**240. Enccladus.**

If Enceladus was really billions of years old we would not expect it to be geologically active (by that I mean the Evolutionist model would predict that Enceladus would be old/cold/dead). However Saturn's moon spews massive jets of water from a south pole fountain into space. What's also incredible is that Enceladus is actually spray painting Saturn's other moons with snow and ice.

"Our Created solar system 1hr 15min" Narrated by Spike Psarris [Youtube]

**241. Titan.**

Let me break this down...

· Titan's atmosphere contains Methane.

· Sunlight breaks Methane down into Ethane.

· Evolutionist predict a global ocean of Ethane on Titan's surface

· *Titan still has Methane, which shouldn't be there if more than a few million of years old.

· It turns out that Titan's surface is dry, with only some inadequate spots of Ethane lake beds.

· Upon discovering around 10% of Titan's surface, there are only 4 craters so far.

"Our Created solar system 1hr 17min" Narrated by Spike Psarris [Youtube]

* That is, without some source of new methane.

## 242. Distant Mature Galaxies.

According to evolutionary dogma the first elements in the universe were hydrogen and helium then over millions of years (or longer) the stars that formed eventually made the heavier elements, such as metals. Why is it then that distant galaxies appear to have just as high as concentrations of these heavy metals as closer galaxies? The distant galaxies appear just as mature as the nearer ones.

https://www.icr.org/article/distant-galaxies-look-too-mature-for

## 243. Spiral arm Galaxies.

The galaxies are spinning, however the stars in the middle are traveling quicker than the ones on the end. The spiral arms shouldn't exist if the billion of years theory holds true.

" Kent Hovind Creation Seminar 1, 1 hr. "

## 244. Supernova remnants.

A super nova is when a star blows up, this happens about every thirty years. Why is there less than 300 of them. This is consistent with a universe that is only thousands of years old. If the universe was billions of years old shouldn't we expect to find much more of them.

" Kent Hovind Creation Seminar 1, 1hr 12min"

## 245. Short Period Comets.

Comets cannot last millions of years, they break apart over time. So why do we still have comets? Evolutionists will tell you that more comets enter our solar system from what is know as a Ooart cloud. There is no evidence for an Ooart cloud, it is not visible and to my knowledge we have no instruments that can detect one.

"Ultimate Proof for Creation/God's Existence"

## 246. Globular Clusters.

According to the evolutionary model, they shouldn't exists. Just in our own milky way there is about 200 of these clusters; each containing 200 thousand to a million stars. Despite their massive size, globular clusters are very stable. This shouldn't be the case if evolution were true, only God could keep these stars from smashing into one another.

The Evolution Handbook by Vance Ferrell pg 99

# What about Aliens?

## 247. Alien ships

Along with the theory of evolution comes the notion that life may have evolved elsewhere in our universe, if the universe is really billions or millions of years old. Some people actually believe this is the case. However, no matter how far this theoretical alien race has evolved, their are some insurmountable hurtles they simple could not overcome. Here are only a few reasons why Aliens cannot be visiting us from other planets.

**1. Cosmic radiation**- Our atmosphere protects us from cosmic rays, people in space; not so much. Additionally as you travel faster, so does the dose of radiation increase.

**2. Physics** tells us that as our speed increase, so does our mass. It has been theorized that as you approach light speed your mass would become infinite.

**3. Space is not empty**, it is filled with dust particles and traveling at just a tenth the speed of light would cause each dust grain to impact your spacecraft with a force equivalent of 10 tonnes of TNT.

**4. Gravitational forces**-Some Air force pilots experience up to 9g for a few seconds when they do those sharp turns, but run the risk of blacking out or dying if trying to do it for longer.

"Alien Intrusion: UFOs and the Evolution Connection-30-11-22" Youtube

## 248. Alien encounters

In order to avoid the distance travel dilemma's many serious UFOlogist believe in (IDH) otherwise known as the interdimesional hypothesis. That is they believe aliens are visiting us from other dimensions. They supposedly have evolved to a higher state so that their bodies differ from our own. This is what the evidence actually implies.

- You don't see aliens arriving from outside the atmosphere.

- UFO's can materialize as well as dematerialize

- Been known to morph and combine into each another

- Ariel maneuvers that are not confined to laws of physics

UFO's have been seen on radar as well as by multiple reliable eyewitnesses.

A christian by the name of Gary Bates (a worker from creation ministries) speaks of the time when another supposed christian calls him and tells her experience where aliens visited her and told her that Jesus was an advanced extraterrestrial like them. It has been claimed that by some that aliens parted the red sea, Elisha was taken up into a mothership, etc.

Similarities of abduction experiences.

The aliens can go through walls and quickly appear in abductee's room. Abductee's have no control during the situation. People have received marks and/or been injured during their experience. There are Similarities to NDE'S. There are no conscious recollection of events.

There is a common sequence of events that occur during abductions (that abductee's remember under hypnosis to try and figure out where that lost time went), there are 8 markers although not everyone always has all eight. The capture, a disturbing examination, a tour of the ship, a conference, a journey to another planet, theopany (meeting a religious person or supposedly divine being), the return back to earth, and aftermath.

Much of the books that discuss UFO'S also deal with the metaphysical and mystical, such as automatic writing, telepathy, and beings that are not visible. Many of the reports of UFO'S currently being published are describing instances that are very much like demonic possession and psychic phenomenon.

Is there intelligent life on other planets? I will so know because mankind has dominion over the earth, and aliens could violate that principle. Plus that would mean aliens would have been punished for Adam's sin and therefore cursed without a saviour because Christ was only offered once for sins and he is our kinsmen redeemer. This would seem to mar the character of God.

Gen 1:28 And God blessed them, and God said unto them, Be fruitful, and multiply, and replenish the earth, and subdue it: and have dominion over the fish of the sea, and over the fowl of the air, and over every living thing that moveth upon the earth.

Rom 8:22 For we know that the whole creation groaneth and travaileth in pain together until now.

Heb 2:11 For both he that sanctifieth and they who are sanctified *are* all of one: for which cause he is not ashamed to call them brethren,

"Alien Intrusion: UFOs and the Evolution Connection-30-11-22" Youtube

# Ch. 6 Philosophical Arguments for the Supreme Mind

## Popular Case for God Arguments

### 249. Laws of Logic

Atheists use laws of logic but they have to borrow it from the christian worldview in order to attack the christian worldview. It is the bible that gives us a foundation for "the preconditions for the intelligibility of man's experience and reasoning", as my source states. In other words we would not be able to prove anything to be true unless the bible was true. The laws of logic is just one example of a precondition of intelligibility. Laws of logic can be defined as the right standard of reasoning, how can there be such a thing in a purposeless random chance world? The laws of logic makes sense under Christianity because they reflect the mind of God.

"Ultimate Proof for Creation/God's Existence 24&26-33min" by Jason Lisle [Youtube]

### 250. Absolute Morality (Conscience)

This one is easy to explain. Without God there can be no objective right or wrong. It would just be your opinion vs mine. However there are objective moral values (such as rape is wrong, torturing babies is wrong, murder is wrong). Therefore God exists. They come from his nature. We all have a conscience given to us by God (Romans 2:14-15). Though some choose to sear it (1 Timothy 4:2).

"Ultimate Proof for Creation/God's Existence 26min" by Jason Lisle [Youtube]

### 251. The Argument for Conscience.

Even people who refuse to admit the truth of objective morality, will still say that it is wrong to violate one's own conscience. Even if that conscience tells people to do or avoid different things, virtually everyone will agree that one should follow their conscience. Whence cometh the authority of the conscience? There are only four possibilities.

· From something less than self i.e. nature. Why would there be an obligation to obey something that is less important than myself.

· Self- If our obligation comes from ourselves can we not then choose to non-obligate ourselves to do or not do a particular thing.

· Others (equal to self)- What right do other have to impose their values on me? Does majority vote decide truth? That is a logical fallacy, Bandwagon(or appeal to popularity).

· God (above self)- The only thing that has a right to demand my obedience would be something superior to self, namely Jesus.

https://www.peterkreeft.com/topics-more/20_arguments-gods-existence.htm

## 252. Kalam Cosmological Argument

The Kalam Cosmological argument is an incredibly powerful argument for God. It goes like this

1. Whatever begins to exists has a cause for it's existence. (I'm sure most would agree with this point).

2. Our universe had a beginning.

· The universe could not have created itself because that would mean that it had to pre-existed before it began to exist, a logical absurdity.

· The universe had to have a beginning because time had a beginning. How do we know time had a beginning. Because if there was an infinite about of past events, we would never reach this point in time where you are reading this. Here is a diagram to help the reader grasp this concept because I know it can be difficult to understand. x=this point in time, 0=infinity. Just as we cant get from X______0 to infinity in the future. So we also cannot get from 0_______X eternity in the past to this point in time. (More on this in a moment.

3. Therefore the universe has a cause.

This is the basics of the Kalam Cosmological Argument. Now of coarse I'm sure we can go deeper than this into the reasoning for what that cause was. The cause of the universe could not be a part of this universe, it had to be space-less, timeless, immaterial, powerful, highly intelligent, changeless, and personal. It had to be a choice maker in order to create an effect in time, or else it wouldn't give rise to anything. The seven attributes that I listed for the cause of the universe are some of the attributes of the God of the bible.

**Space-less**- Jer 23:24 Can any hide himself in secret places that I shall not see him? saith the LORD. Do not I fill heaven and earth? saith the LORD.

**Timeless**- Psa 90:2 Before the mountains were brought forth, or ever thou hadst formed the earth and the world, even from everlasting to everlasting, thou *art* God.

**Immaterial**- Joh 4:24 God *is* a Spirit: and they that worship him must worship *him* in spirit and in truth.

**Powerful**- Rev 19:6 And I heard as it were the voice of a great multitude, and as the voice of many waters, and as the voice of mighty thunderings, saying, Alleluia: for the Lord God omnipotent reigneth.

**Intelligent**- Psa 147:5 Great *is* our Lord, and of great power: his understanding *is* infinite.

**Changeless**- Mal 3:6 For I *am* the LORD, I change not; therefore ye sons of Jacob are not consumed.

**Personal**-Rev 3:20 Behold, I stand at the door, and knock: if any man hear my voice, and open the door, I will come in to him, and will sup with him, and he with me.

https://en.wikipedia.org/wiki/Kalam_cosmological_argument

"Proving God Exists-Scientifically, 3superdogs"

## 253. Ontological Argument

The argument can be summed up as:

a) If it is possible God exist, God exists.

b) It is possible God exists.

c) Therefore God exists.

https://www.cedarville.edu/insights/post/what-is-the-ontological-argument

## 254. Law of Information

"There is no known law of nature, no known process and no known sequence of events which can cause information to originate by itself in matter." Dr. Werner Gitt.

"Ultimate Proof for Creation/God's Existence by Jason Lisle 1 min" [YouTube]

## 255. 2nd Law of Thermodynamics

Simply speaking, everything physical breaks down with time. Cars rust, Paints fade, Buildings crumble. The earth waxes old (Hebrews 1:11-12). This refutes the idea of an eternal universe. Even living species are getting weaker over time. As I previously showed in other sections we are getting smaller, and in some ways dumber than our ancestors.

" Kent Hovind Creation Seminar 1" supports this teaching.

## 256. Sensus Divinitatus.

Another way to put it is sense of the divine, which is the God-shaped void in our hearts. We all were designed to worship. However not everyone worships the true God. Many worship graven images of false gods, some worship money and makes that their life pursuit. Still others worship rock band artist or football teams. Some even put their own spouse in the #1 place in their heart. We must all guard against Idolatry.

https://credohouse.org/blog/ten-arguments-for-the-existence-of-god

**257. Logical fallacies.**

For this section I will list common and/or interesting logical fallacies that may come up when debating evolution when discussing origins. To keep this section reasonably short I will just give ten examples, but I encourage my readers to study this topic further as logical fallacies are an interesting and important topic in regards to the creation/evolution debate. I will not include the circular reasoning, Bandwagon, appeal to authority or loaded question fallacies as they are mentioned elsewhere in this book.

1. **Ad Homenin**- Attacking the person instead of the argument. Name-calling, reputation smearing, etc.
2. **False Dilemma**- This occurs when one incorrectly limits the possible explanations of something. Evolutionists will never admit to a supernatural cause to an event. Also they will rule out any evidence of dinosaurs living with man.
3. **Cherry picking**- Selecting a certain data set or group of facts and dismissing others in order to frame an argument to suit a specific purpose. Haeckle's Embyo's are a prime example.
4. **Equivocation**- This fallacy is committed when one switches from one meaning of a word to another meaning in an argument. Evolution can mean change (many creationists believe in this factual definition). However evolution can also mean change from one kind to another (macro-evolution), which is unproven and religious in nature.
5. **Appeal to Probability**- Evolutionists arguably commit this fallacy when they speak of the first cell arising out of the pre-biotic soup. Saying that given enough time (no matter how unlikely), life will arise from non-living matter. The appeal to probability is when someone says that just because something is possible (i.e life arising by chance from soup), that it is probable or certain. Ever hear someone claim evolution is a FACT, this is an appeal to probability.
6. **Genetic Fallacy**- Claiming the argument is terrible because of who said it (i.e. the source). "you don't celebrate Easter? Are you a Jehovah's

Witness?"

7.  **Appeal to Pity**- Supporting your argument by relying on feelings and sympathy " God's not real cuz, starving children in Africa". Indeed the problem of evil is one of the biggest and most popular arguments used to attack the idea of a belief in a God. What some fail to realize is that without free will we are not capable of loving anyone, but with God giving us free will, this also opened up the possibility of us committing evil. God didn't want robots. We are to blame for the suffering in the world by misusing our free will. Also you can't argue against God by using evil as an argument. Without God there is no standard for good and evil.

8.  **Tu Quoque**- The you too argument. This is when you attack the behavior of a person on the count of they're being inconsistent with their beliefs, aka hypocrisy. Christian's are frequently blamed for the inquisition, yet this doesn't invalidate the truth of Christianity. For one the inquisition was a Catholic endeavor (not true biblical Christianity), also the bible warns of wolves in sheep's clothing.

9.  **Strawman**- Building and then attacking an inferior version of the opponents argument. "How did Noah's get all those millions of animals on that itty bitty boat? Umm...the boat was huge, there were not millions of animals, probably only a few or several thousand. According to Kent Hovind's seminar 3 there are only about 8,000 kinds in the world (around 9 min mark).

10. **No True Scottsman Fallacy**- Occurs when one changes a definition of a word in order to support their argument. Example: Evolutionist says "No Scientists rejects evolution". Creationists replies "Dr. Jason Lisle is a scientists and he rejects evolution". Evolutionists replies "No ***true scientists*** rejects evolution". The evolutionists has committed the No true scottsman fallacy.

## 258. Aesthetic Experience.

This is arguing from universal beauty and/or pleasure.

https://credohouse.org/blog/ten-arguments-for-the-existence-of-god

## 259. Teleological Argument

The Teleological argument is when one points to the appearance of design and or purpose of nature as evidence of a designer. Feel free to see the link below for a more detailed description. Throughout this book I have listed many instances of Teleological arguments. From the complexity of DNA, to complex chemicals such as Enzymes & Proteins, I have also pointed to living organisms such as the cell, the giraffe, the woodpecker, cuttlefish, the rat, the bird, and many other organisms to show how they refute the notion of random mutations producing such beings. Saturn's dancing moons and our Eclipse may be considered Teleological arguments.

https://www.gotquestions.org/teleological-argument.html

## 260. The Golden Ratio.

Or Fibonacci sequence goes like this 0,0,1,2,3,5,8,13,21,34,55, and so on. Each number is the sum of the previous two parts. It can be seen in many things such as shells, flower pedals, pine cones, spiral galaxies, certain animals bodies, as well as human faces, just to name a few.

https://www.mathnasium.com/blog/14-interesting-examples-of-the-golden-ratio-in-nature

## 261. Paschel's Wager.

This basically states that due to the consequences of being wrong, belief in God is the most rational choice. For if a Christian is wrong he lived a good life and loses nothing in the end. But if the Atheists is wrong, eternal suffering is their portion.

https://credohouse.org/blog/ten-arguments-for-the-existence-of-god

# Other Points to Consider

## 262. Science is dependent on the Creator

Can't do science without reliable memories and trustworthy brain...but evolution says our thoughts are just molecules bouncing and fizzing around in our skulls, no freewill.

## 263. Theories about the age of earth have varied widely, man is fallible.

Up until the 1700's most people believed in a young earth, while the age of the earth based on the bible's timeline has been quite consistent (Approx 4000 B.C; give or take 1000 years), the timeline given by secular and old earth proponents have changed drastically over the last few hundred years. Have a look at the enormous difference in ages given in just a short amount of time. Note: these ages are not scientific at all, as we talked about earlier in order to date fossils or rocks you have to make some assumptions which makes the dates fallible.

- Comte de Buffon, 78000 years old. 1779 A.D.

- Abraham Werner, 1 Million years. 1786 A.D.

- Charles Lyell, Millions of years. 1830's.

- Arthur Holmes, 1.6 Billion years ago. 1913 A.D.

- Clair Patterson, 4.5 Billion. 1956 A.D.

https://answersingenesis.org/age-of-the-earth/how-old-is-the-earth/

## 264. Bible explains why we have 7 day week, evolution does not.

## 265. Why we are ashamed by being naked, explained by bible not evolution.

## 266. The Moral superiority of Christianity

When contrasted to other religions Christianity has the superior moral ground. Though many religions have some shared values (perhaps due to the fact we all have an inborn knowledge of right from wrong), only Christianity agrees 100% with our inborn conscience. This is because our conscience was given by God. Sadly, I will not be giving a breakdown of a bunch of the problems associated with other faiths. I will leave the reader to do their own research in this regard.

# Ch. 7 More Evidence for Christianity Sure to Delight.

## What about the Supernatural

### 267. Near Death Experiences

One big problem with evolution is that is completely discounts personal testimonies of NDE (Near-Death-Experiences), there is a channel on Youtube called **'Heaven Awaits'**, it contains many video's of documented NDE's. From an evolutionary worldview how could we explain such instances of experiences of those people who have died briefly and then came back to life? Sure, some of them could be fake, caused by dishonest people out for fame, but *all* of them? If I had to assume, I would guess that there are at least thousands of such cases if not much more. Could random chemicals in a dying brain cause someone to imagine such vivid experiences that they have mistaken for out of body experiences? I don't think so. Even as a Christian I have some skepticism of those who have claimed to have seen heaven in light of some scriptures, this doesn't mean though that I disbelieve hell testimonies though. Here are a few verses that may make one skeptical of those who have claimed to have seen heaven.

> · Joh 3:13 And no man hath ascended up to heaven, but he that came down from heaven, *even* the Son of man which is in heaven.

> · 1Co 2:9 But as it is written, Eye hath not seen, nor ear heard, neither have entered into the heart of man, the things which God hath prepared for them that love him.

There is a book called '23 Minutes in Hell' by Bill Weise, it is a very interesting read. The only thing in the book that I could tell might be extra biblical is the account of him being tortured by demons in hell. Though there may be a few verses that back up what he was saying. In any case I recommend reading it, it does contain many verses on hell.

## 268. Personal Miracle Testimonies & Healing.

What are evolutionists to say when someone tells them of miraculous occurrences in their life? The evolutionist has only a few options. They can say that they are lying, mistaken, or they can believe them. Do miracles happen today? The answer is a resounding Yes! There really is no excuse for not believing in the supernatural when it is all around us. Those who have been in the church a long time know that the Lord Jesus is still working in peoples lives today. Now are there fraudsters out there and fake faith healers, no doubt, but this doesn't negate the legitimacy of other people stories of miracles and healing. For the remainder of this section I will list a few examples from miracles and healings that my relatives and wife has experienced, then I will give a couple of relevant channels for my reader to do more research on this topic.

My wife Sarah has two separate occasions were she was healed. The first time was when she was a teenager and her parents had the elders of the church pray over her and Jesus heard their prayer of faith, Sarah never had a seizure again. That was over 15 years ago. The next healing was after she had married me and we were staying at my grandmothers house. While I was away working, my wife was sick with the swine flu, at the prompting of a TV preacher (who was specifically addressing those with her illness), she came into with his prayer and was immediately healed (like within 1 or 2 minutes).

There have been many instances of my mother in law entertaining guest, and during those financial hard time the Lord literally multiplied their food. She would just not look into the pot and with faith kept scooping out of the container a good serving for each of her guest, and there was always enough, more than there should have been under natural conditions. Another one of my Relatives used his propane for about a year and when the time came for the refill, the service man claimed that it was already full.

Youtube Channels that contain or may contain testimonies of Supernatural Events

1.  Sid Roth's It's Supernatural!
2.  The 700 Club and 700 Club Interactive
3.  Richard Loranzo Jr.

## 269. Magicians.

Now I know this is not the best piece of evidence to give an evolutionists to convince them that their theory is false. Mainly because we live in an age of CGI and just about anything you see on the internet can be fake. However some of the strange things performed by magicians are done in front of an audience so unless they are just going along and pretending to be amazed for the camera, some of the stuff is kinda hard to fake. Now I do not want my readers watching magicians on youtube so I will not name names of magicians. I will however describe some of the things that I have seen done by magicians. Now I am aware that probably most of what we see on television is sleight of hand, trap doors, props, etc. However, according to the bible sorcery is real and is not to me meddled with. I do believe there are real magicians out there though who are truly in league with evil spirits, woe unto them! Here is a short list of what I've seen them do.

1.  Levitation
2.  Teleportation-like stuff
3.  Impaling themselves with sword or lancets, while feeling no harm
4.  Putting large objects into glass bottles
5.  Reaching through glass
6.  One guy being cut in half, then riding a unicyle before being put back togther
7.  Card tricks
8.  Manipulating fire
9.  Mask changing

## 270. Ghost.

Another problem that I have with evolution is that it completely dismisses the idea of ghosts. In evolutionary theory, matter is all there is. Now even I don't believe departed loved ones are free to roam the earth as they please according to my worldview, however I do believe in angels and devils. They can certainly be mistaken for a ghost. I cannot write off "ghosts" sightings because I have personally known at least 5 people that have seen a ghost. In one instance two people saw a ghost at the same time, that make writing it off as an illusion very difficult. Besides there are innumerable more ghost sightings worldwide, are we to expect that there are natural explanations for all of them. What is the possibility of that?

## 271. Supernatural occurrences like Prophesy, Visions, Deliverance, and Spiritual gifts.

The Naturalistic worldview (Evolutionary theory) has no framework for supernatural occurrences. According to an evolutionists things like healings, miracles, visions, and prophesy are all illegitimate. According to atheists these people are just lying or are mistaken, or delusional, or crazy, etc. But does this fit the evidence? The answer is a resounding no. How do you fake a word of knowledge about someone. A word of knowledge is when God told you something about someone that there would have been no way you would have known about otherwise. An example of a word of knowledge would be a fellow believer pulling you to the side and telling you that God saw that you were watching that XXX rated movie last night and you'd better repent. There would be no other explanation of how that Christian would know this apart from a word of knowledge.

Furthermore how do we explain visions. We cannot chalk it up to random chemical reactions in the brain for we know information always comes from a mind. A vision is an ordered, coherent, "film" as it were that displays information. Sure it may contain symbolism but at least they make sense. It is unreasonable to say someone is lying if they claim to have a vision (especially, if they have a history of otherwise being a rational, and trustworthy person). Also how do we explain the testimonies of those who have been healed as a result of the prayers of the faithful, (but the Lord Jesus is the one doing the healing). Yes I'm aware that there are fake healers out there but the counterfeit does not negate the fact of genuine healing.

If you want examples of the Spiritual Gifts in use, I recommend checking out the following youtube channels. (Note: It is the conviction of this author that not all cases of speaking in tongues is legitimate, many well meaning Charismatics/Pentecostals have been trained to speak this "heavenly prayer language" that I believe is not biblical). With that aside here is a list of Seven resources for the reader to check out.

1. Isaiah Salvidor (deliverance)
2. Richard Lorenzo jr.
3. Greg Locke
4. Heshkinah
5. John Ramirez
6. Nicholas Bowling
7. Troy Black

Needless to say I may not agree with 100% of everything these teaches say but I do enjoy watching them. I do believe that these are strong leaders and/or believers in the Christian community today.

# It is Evolution, Not the Bible That's Anti-Science

### 272. Not all Scientists are Evolutionists.

For this section I will give just a small sample of Scientists who believed in God. I will include ten historical figures and five of my personal favorites of modern scientists. For a much greater list (dozens, if not hundreds of examples), consider the website below.

1. Francis Bacon (1561–1626) The Scientific method.

2. Galileo Galilei (1564–1642) Astronomy & Physics

3. Robert Boyle (1627–1691) Chemistry; Gas dynamics.

4. Isaac Newton (1642–1727) Reflecting telescope; Dynamics; Gravitation law; Spectrum of light; Calculus.

5. John Dalton (1766–1844) Father of the Modern Atomic Theory; Chemistry.

6. Peter Mark Roget (1779–1869) Physician; Physiologist

7. Henry Rogers (1808–1866) Geology

8. Louis Pasteur (1822–1895) Sterilization; Immunization; Chemical chirality, Bacteriology, and Biochemistry;

9. Ferdinand von Mueller (1825–1896) Botanist.

10. Alexander MacAlister (1844–1919) Anatomy

11. Dr Jack W. Cuozzo (1937–2017) Dentist

12. Dr Henry M. Morris (1918–2006) Hydrologist

13. Dr Steve Austin, Geologist

14. Dr Jason Lisle, Astrophysicist

15. Dr Andrew Snelling, Geologist

creation.com/creation-scientists

## 273. Forty Quotes From Scientists and Evolutionists on Evolution.

"The fossil record with its abrupt transitions offers no support for gradual change..." Dr. Stephen Jay Gould "The Return of Hopeful Monsters,' *Natural History*, 86[4]:22-30, June-July, 1977.

"Evolution itself is accepted by zoologists, not because it has been observed to occur ... or can be proved by logical coherent evidence, but because the only alternative, special creation, is clearly incredible." D. M. S. Watson, "Adaptation," Nature, Vol. 123, p. 233 (1929).

"It is our knowledge of how these organisms actually operate, not speculations about how they may have arisen millions of years ago, that is essential to doctors, veterinarians, farmers ... "

– Skell, P.S., The Dangers Of Overselling Evolution; Focusing on Darwin and his theory doesn't further scientific progress, Forbes magazine, 23 Feb 2009

"A commonsense interpretation of the facts suggests that a superintellect has monkeyed with physics, as well as chemistry and biology, and that there are no blind forces worth speaking about in nature." Sir Frederick Hoyle, "The Universe: Past and Present Reflections." Engineering and Science, November, 1981. pp. 8–12

"Evolution is unproved and unprovable. We believe it only because the only alternative is special creation, and that is unthinkable." Sir Arthur Keith, Criswell, W.A. (1972), Did Man Just Happen?, Grand Rapids, MI, Zondervan, p. 73.

"To postulate that the development and survival of the fittest is entirely a consequence of chance mutations seems to me a hypothesis based on no evidence and irreconcilable with the facts. These classical evolutionary theories are a gross over-simplification of an immensely complex and intricate mass of facts, and it amazes me that they are swallowed so uncritically and readily, and for such a long time, by so many scientists without murmur of protest." (Sir Ernest Chain, "Was Darwin Wrong" by Francis Hitching, LIFE magazine, April 1982, p.50)

"Even if all the data point to an intelligent designer, such an hypothesis is excluded from science because it is not naturalistic." Dr. Scott Todd, Kansas State University, Nature 401(6752):423, Sept. 30, 1999.

The probability of life originating from accident is comparable to the probability of the unabridged dictionary resulting from an explosion in a printing shop – Dr. Edwin Conklin, (1986) Give Me an Answer, p. 70

"There are only two possibilities as to how life arose; one is spontaneous generation arising to evolution, the other is a supernatural creative act of God, there is no third possibility. Spontaneous generation that life arose from non-living matter was scientifically disproved 120 years ago by Louis Pasteur and others. That leaves us with only one possible conclusion, that life arose as a creative act of God. I will not accept that philosophically because I do not want to believe in God, therefore I choose to believe in that which I know is scientifically impossible, spontaneous -generation arising to evolution." – Dr. George Wald, Scientific American, August,1954 Origin of Life article, p.48

"My attempts to demonstrate evolution by an experiment carried on for more than 40 years have completely failed.....It is not even possible to make a caricature of an evolution out of paleobiological facts...The idea of an evolution rests on pure belief." – Dr. Nils Heribert-Nilsson, Synthetische Artbildung

"Transformism is a fairy tale for adults." (Age Nouveau, [a French periodical] February 1959, p. 12).

"In conclusion, evolution is not observable, repeatable, or refutable, and thus does not qualify as either a scientific fact or theory." – Dr. David N. Menton, PhD , Is evolution a theory, a Fact, or a Law?— or None of the Above?

"The chance that higher life forms might have emerged through evolutionary processes is comparable with the chance that a tornado sweeping through a junk yard might assemble a Boeing 747 from the material therein." – Sir Fred Hoyle, Radio Lecture 1982

"The probability of dust carried by the wind reproducing Durer's 'Melancholia' is less infinitesimal than the probability of copy errors in the DNA molecule leading to the formation of the eye; besides, these errors had no relationship whatsoever with the function that the eye would have to perform or was starting to perform. There is no law against daydreaming, but science must not indulge in it." Dr Pierre-Paul de Grasse, Zoologist. 'Evolution of Living Organisms' 1977, p104.

"Stasis, or non-change, of most fossil species during their lengthy geological life spans was tacitly acknowledged by all paleontologists, but almost never studied explicitly because prevailing theory treated stasis as uninteresting non-evidence for non-evolution. ... The overwhelming prevalence of stasis became an embarrassing feature of the fossil record, best left ignored as a manifestation of nothing (that is, non-evolution)." Gould, Stephen J., "Cordelia's Dilemma," Natural History, 1993, p. 15.

"Darwinian theory is the creation myth of our culture. It's the officially sponsored, government financed creation myth that the public is supposed to believe in, and that creates the evolutionary scientists as the priesthood... So we have the priesthood of naturalism, which has great cultural authority, and of course has to protect its mystery that gives it that authority—that's why they're so vicious towards critics." Phillip Johnson, On the PBS documentary "In the Beginning: The Creationist Controversy" [May 1995]

"We are told dogmatically that Evolution is an established fact; but we are never told who has established it, and by what means. We are told, often enough, that the doctrine is founded upon evidence, and that indeed this evidence 'is henceforward above all verification, as well as being immune from any subsequent contradiction by experience;' but we are left entirely in the dark on the crucial question wherein, precisely, this evidence consists." Smith, Wolfgang (1988) Teilhardism and the New Religion: A Thorough Analysis of The Teachings of Pierre Teilhard de Chardin Rockford, Illinois: Tan Books & Publishers Inc., p.2

"The point, however, is that the doctrine of evolution has swept the world, not on the strength of its scientific merits, but precisely in its capacity as a Gnostic myth. It affirms, in effect, that living beings created themselves, which is, in essence, a metaphysical claim. ... Thus, in the final analysis, evolutionism is in truth a metaphysical doctrine decked out in scientific garb" (*Teilhardism and the New Religion*, p. 24). Wolfgang Smith, Ph.D.

"Scientists who go about teaching that Evolution is a fact of life are great con men, and the story they are telling may be the greatest hoax ever. In explaining Evolution we do not have one iota of fact." (Dr T N Tahmisian, Atomic Energy Commission, The Fresno Bee, August 20, 1959.

"Evolutionism is a fairy tale for grown-ups. This theory has helped nothing in the progress of science. It is useless," Professor Louis Bouroune, The Advocate, March 8, 1984.

"The curious thing is that there is a consistency about the fossil gaps; the fossils are missing in all the important places." Francis Hitching, The Neck of the Giraffe or Where Darwin Went Wrong, Penguin Books, 1982, p.19)

"It is possible (and, given the Flood, probable) that materials which give radiocarbon dates of tens of thousands of radiocarbon years could have true ages of many fewer calendar years." Gerald Aardsma,

https://www.icr.org/article/myths-regarding-radiocarbon-dating

"Evolution requires intermediate forms between species and palaeontology does not provide them." David Kitts, Paleontology and evolutionary theory, *Evolution* **28**:467, September 1974.

"In any case, no real evolutionist, whether gradualist or punctuationist, uses the fossil record as evidence in favour of the theory of evolution as opposed to special creation." Ridley, Mark, "Who doubts evolution?" New Scientist, vol. 90, 25 June 1981, p. 831.

"Since the fossil material provides no evidence of other aspects of the transformation from fish to tetrapod, paleontologists have had to speculate how legs and aerial breathing evolved." Stahl, Barbara. 1974. Vertebrate history: Problems in evolution, New York: Dover Publications, Inc. p. 195

"To propose and argue that mutations even in tandem with 'natural selection' are the root-causes for 6,000,000 viable, enormously complex species, is to mock logic, deny the weight of evidence, and reject the fundamentals of mathematical probability." Cohen, I.L. (1984) Darwin Was Wrong: A Study in Probabilities , New York: New Research Publications, Inc., p. 81

"There are not enough fossil records to answer when, where, and how Homo sapiens emerged." Takahata, Molecular Anthropology, Annual Review of Ecology & Systematics, 1995, p. 355

Evolution is promoted by its practitioners as more than mere science. Evolution is promulgated as an ideology, as secular religion—a full-fledged alternative to Christianity, with meaning and morality. I am an ardent evolutionist and an ex-Christian, but I must admit that in this one complaint—and Mr. Gish is but one of many to make it—the literalists are absolutely right. Evolution is a religion. This was true of evolution in the beginning, and it is true of evolution still today" Ruse, M. (2000). National Post.

"The extreme rarity of transitional forms in the fossil record persists as the trade secret of paleontology. The evolutionary trees that adorn our textbooks have data only at the tips and nodes of the branches; the rest is inference, however reasonable, not the evidence of fossils." —Stephen J. Gould, Natural History, Volume 86, Number 5, May 1977 (p. 14)

". . . the most famous example . . . still on exhibit downstairs is the exhibit on horse evolution. . . That has been presented as literal truth in textbook after textbook. . . [T]he people who propose these kinds of stories themselves may be aware of the speculative nature of some of the stuff. But by the time it filters down to the textbooks, we've got science as truth and we've got a problem."
——Dr. Niles Eldridge

"Archaeopteryx probably cannot tell us much about the early origins of feathers and flight in true protobirds because Archaeopteryx was, in a modern sense, a bird." Alan Feduccia, 1993 Science Article.

"...there seems to have been almost no change in any part we can compare between the living organism and its fossilized progenitors of the remote geological past. Living fossils embody the theme of evolutionary stability to an extreme degree. ....We have not completely solved the riddle of living fossils." Niles Eldridge, his book FOSSILS 1991, p. 101, 108.

"Our theory of evolution has become, as Popper described, one which cannot be refuted by any possible observations. Every conceivable observation can be fitted into it. It is thus 'outside of empirical science but not necessarily false'. No one can think of ways in which to test it." Charles Birch and Paul Elrlich, Evolutionary history and population biology, Nature 214:352, 1967.

"Scientism is not the same thing as science. Science is a blessing, but scientism is a curse. Science, I mean what practicing scientists actually do, is acutely and admirably aware of its limits, and humbly admits to the provisional character of its conclusions; but scientism is dogmatic, and peddles certainties. It is always at the ready with the solution to every problem, because it believes that the solution to every problem is a scientific one, and so it gives scientific answers to non-scientific questions. Owing to its preference for totalistic explanation, scientism transforms science into an ideology, which is of course a betrayal of the experimental and empirical spirit." Leon Wieseltier, Perhaps Culture is Now the Counterculture: A Defense of the Humanities, 19 May 2013; www.newrepublic.com/article/113299/leon-wieseltier-commencement-speech-brandeis-university-2013[1]

"There is a popular image of human evolution that you'll find all over the place ... On the left of the picture there's an ape ... On the right, a man ... Between the two is a succession of figures that become ever more like humans ... Our progress from ape to human looks so smooth, so tidy. It's such a beguiling image that even the experts are loath to let it go. But it is an illusion." Bernard Wood, "Who are we?" New Scientist, 2366 (October 26, 2002): 44.

"The problem that biological evolution poses for natural theologians is the sort of God that a darwinian version of evolution implies ... The evolutionary process is rife with happenstance, contingency, incredible waste, death, pain and horror ... Whatever the God implied by evolutionary theory and the data of natural history may be like, He is not the Protestant God of waste not, want not. He is also not a loving God who cares about His productions. He is not even the awful God portrayed in the book of Job. The God of the Galápagos is careless, wasteful, indifferent, almost diabolical. He is certainly not the sort of God to whom anyone would be inclined to pray." David L. Hull, "The God of the Galápagos," review of Phillip Johnson's Darwin on Trial, Nature, 352 (August 8, 1991): 485–6, https://doi.org/10.1038/352485a0.

"For use in understanding the evolution of vertebrate flight, the early record of pterosaurs and bats is disappointing: Their most primitive representatives are

---

1. http://www.newrepublic.com/article/113299/leon-wieseltier-commencement-speech-brandeis-university-2013

fully transformed as capable fliers." Sereno, Paul C., The evolution of dinosaurs, Science 284(5423):2137–2147 (p. 2143), June 25, 1999.

'I have faith and belief myself. I believe that the universe is comprehensible within the bounds of natural law and that the human brain can discover those natural laws and comprehend the universe. I believe that nothing beyond those natural laws is needed.'

'I have no evidence for this. It is simply what I have faith in and what I believe.'

Isaac Asimov, Counting the Eons, Grafton Books (Collins), London, p.10.

"Let's be clear: the work of science has nothing whatever to do with consensus. Consensus is the business of politics. Science, on the contrary, requires only one investigator who happens to be right, which means that he or she has results that are verifiable by reference to the real world. In science consensus is irrelevant. What is relevant is reproducible results." Michael Crichton, Caltech lecture, 2003

"Practically all orders and families known appear suddenly and without any apparent transitions."

Dr. Richard B. Goldschmidt, "Evolution as Viewed by One Geneticist," American Scientist, January 1952, p. 97.

Quotes #1, #3-#6, #8-#10, #12-#14

https://bsssb-llc.com/quotations-from-scientists-evolutionists-about-darwinian-evolution/#:~:text=The%20theory%20of%20evolution%20may,accord%20with%20its%20basic%20idea[2].

Quotes #2, #7, #15, #24, #25, #27, and #40

https://www.creationworldview.org/evolution-is-a-religion

Quote #11

https://creationdesign.org/english/commentsonevolution.html

---

2. https://bsssb-llc.com/quotations-from-scientists-evolutionists-about-darwinian-evolution/#_853ae90f0351324bd73ea615e6487517__4c761f170e016836ff84498202b99827__853ae90f0351324bd73ea615e6487517_text_43ec3e5dee6e706af7766fffea512721_The_0bcef9c45bd8a48eda1b26eb0c61c869_20theory_0bcef9c45bd8a48eda1b26eb0c61c869_20of_0bcef9c45bd8a48eda1b26eb0c61c869_20evolution_0bcef9c45bd8a48eda1b26eb0c61c869_20may_c0cb5f0fcf239ab3d9c1fcd31fff1efc_accord_0bcef9c45bd8a48eda1b26eb0c61c869_20with_0bcef9c45bd8a48eda1b26eb0c61c869_20its_0bcef9c45bd8a48eda1b26eb0c61c869_20basic_0bcef9c45bd8a48eda1b26eb0c61c869_20idea

Quotes #16, #17, #21-#23, & #26

https://www.rae.org/essay-links/quotes/

Quote #18

https://www.wayoflife.org/reports/evolutionists_against_darwinism.html

Quotes #19-#20

https://www.chick.com/battle-cry/article?id=scientists-admit-evolution-not-supported-by-facts!

Quotes #28

https://creationtoday.org/scientists-quotes-about-evolution/

Quotes #29-#32

https://www.cs.unc.edu/~plaisted/ce/challenge9.html

for the sake of space the following is a link for the page that contains links for #'s 33-39.

https://creation.com/quotable-quotes

Reader Please note that although the links provided contain the quotes and the person quoted, some of the details of this section (such as work name, year published and pg number of the quotes, or website link) may have been gathered from other sources from the web.

## 274. Bible scientific Foreknowledge.

According to Evolution mankind descended from either monkey/apes or shared a common ancestor with them. Evolutionists believe man started off stupid and become more civilized and smart with time. But is this true? We have already taken a look at some ancient technology which refutes this idea, and here is another, bible scientific foreknowledge.

Throughout the bible you can find scientific principles that ancient people would not have known if evolution was true. This is because mankind was supposedly more primitive and wouldn't have known these things because they were too dumb. But if this was the case why are these scientific facts in bible.

Either mankind was a lot smarter in the past (contrary to evolutionary theory)or it was given by divine revelation (hence also undermining evolution). Here are **39** of such scientific principles and feel free to see 50+ more in the provided link. Note: most but not all of these examples are derived from the following link.

**1. Time had a beginning.**

Gen 1:1 In the beginning God created the heaven and the earth.

**2. 1st Law of Thermodynamics.**

Gen 2:1 Thus the heavens and the earth were finished, and all the host of them.

**3.Man has 28 base and trace elements in common with the earth.**

Gen 2:7 And the LORD God formed man *of* the dust of the ground, and breathed into his nostrils the breath of life; and man became a living soul.

**4. Amazing Sea-faring Vessel.**

Gen 6:15 And this is the fashion which thou shalt make it of: The length of the ark shall be three hundred cubits, the breadth of it fifty cubits, and the height of it thirty cubits.

**5.Rest one day in Seven**

Exo 20:9 Six days shalt thou labour, and do all thy work:

Exo 20:10 But the seventh day is the sabbath of the LORD thy God: in it thou shalt not do any work, thou, nor thy son, nor thy daughter, thy manservant, nor thy maidservant, nor thy cattle, nor thy stranger that is within thy gates:

**6.Letting land rest one in Seven years controls pest and replenishes soil.**

Exo_23:11 But the seventh year thou shalt let it rest and lie still; that the poor of thy people may eat: and what they leave the beasts of the field shall eat. In like manner thou shalt deal with thy vineyard, and with thy oliveyard.

**7.Milk blocks Iron Absorption.**

Exo_34:26 The first of the firstfruits of thy land thou shalt bring unto the house of the LORD thy God. Thou shalt not seethe a kid in his mother's milk.

## 8.Eating Pork bad for Health (pork-worm).

Lev 11:7 And the swine, though he divide the hoof, and be clovenfooted, yet he cheweth not the cud; he is unclean to you.

Lev 11:8 Of their flesh shall ye not eat, and their carcase shall ye not touch; they are unclean to you.

## 9.Quarantining the Sick.

Lev 13:46 All the days wherein the plague shall be in him he shall be defiled; he is unclean: he shall dwell alone; without the camp shall his habitation be.

## 10. How to properly deal with mold.

Lev 14:43 And if the plague come again, and break out in the house, after that he hath taken away the stones, and after he hath scraped the house, and after it is plaistered;

Lev 14:44 Then the priest shall come and look, and, behold, if the plague be spread in the house, it is a fretting leprosy in the house: it is unclean.

Lev 14:45 And he shall break down the house, the stones of it, and the timber thereof, and all the morter of the house; and he shall carry them forth out of the city into an unclean place.

## 11.Good Hygiene.

Lev_15:13 And when he that hath an issue is cleansed of his issue; then he shall number to himself seven days for his cleansing, and wash his clothes, and bathe his flesh in running water, and shall be clean.

## 12. Blood necessary for Life.

Lev 17:11 For the life of the flesh is in the blood: and I have given it to you upon the altar to make an atonement for your souls: for it is the blood that maketh an atonement for the soul.

## 13.Mixed Fabric bad for Health

Deu 22:11 Thou shalt not wear a garment of divers sorts, *as* of woollen and linen together.

## 14.Not Covering Wastes caused black plague.

Deu 23:13 And thou shalt have a paddle upon thy weapon; and it shall be, when thou wilt ease thyself abroad, thou shalt dig therewith, and shalt turn back and cover that which cometh from thee:

## 15.Not Turtle, Not Atlas. Earth's flee float in space.

Job 26:7 He stretcheth out the north over the empty place, and hangeth the earth upon nothing.

## 16.Air has weight.

Job_28:25 To make the weight for the winds; and he weigheth the waters by measure.

## 17.Earth's Rotation.

Job 38:12 Hast thou commanded the morning since thy days; and caused the dayspring to know his place;

Job 38:13 That it might take hold of the ends of the earth, that the wicked might be shaken out of it?

Job 38:14 It is turned as clay to the seal; and they stand as a garment.

## 18.Sea Springs. Did they have submarines back then? or divine revelation.

Job_38:16 Hast thou entered into the springs of the sea? or hast thou walked in the search of the depth?

## 19. Light travels, Dark is stationary.

Job 38:19 Where is the way where light dwelleth? and as for darkness, where is the place thereof,

## 20. Light has a spectrum.

Job 38:24 By what way is the light parted, which scattereth the east wind upon the earth?

## 21. Added this because it reminds me of telephone poles.

Job 38:35 Canst thou send lightnings, that they may go, and say unto thee, Here we are?

## 22.The sea has paths.

Psa 8:8 The fowl of the air, and the fish of the sea, and whatsoever passeth through the paths of the seas.

## 23. The Sun is not stationary.

Psa 19:4 Their line is gone out through all the earth, and their words to the end of the world. In them hath he set a tabernacle for the sun,

Psa 19:5 Which is as a bridegroom coming out of his chamber, and rejoiceth as a strong man to run a race.

Psa 19:6 His going forth is from the end of the heaven, and his circuit unto the ends of it: and there is nothing hid from the heat thereof.

## 24. # of Starts finite, they are different (see point 223), comparable to sand on seashore.

Psa_147:4 He telleth the number of the stars; he calleth them all by their names.

1Co 15:41 There is one glory of the sun, and another glory of the moon, and another glory of the stars: for one star differeth from another star in glory.

Heb_11:12 Therefore sprang there even of one, and him as good as dead, so many as the stars of the sky in multitude, and as the sand which is by the sea shore innumerable.

## 25. Depression Harmful.

Pro_17:22 A merry heart doeth good like a medicine: but a broken spirit drieth the bones.

## 26. Verbal abuse is a Real thing.

Pro_18:21 Death and life are in the power of the tongue: and they that love it shall eat the fruit thereof.

## 27. Air has Circuits.

Ecc 1:6 The wind goeth toward the south, and turneth about unto the north; it whirleth about continually, and the wind returneth again according to his circuits.

## 28. Hydro-logical Cycle.

Ecc 1:7 All the rivers run into the sea; yet the sea is not full; unto the place from whence the rivers come, thither they return again.

## 29. Earth is not a flat rectangle.

Isa_40:22 It is he that sitteth upon the circle of the earth, and the inhabitants thereof are as grasshoppers; that stretcheth out the heavens as a curtain, and spreadeth them out as a tent to dwell in:

## 30. Plants used for Medicine.

Eze_47:12 And by the river upon the bank thereof, on this side and on that side, shall grow all trees for meat, whose leaf shall not fade, neither shall the fruit thereof be consumed: it shall bring forth new fruit according to his months, because their waters they issued out of the sanctuary: and the fruit thereof shall be for meat, and the leaf thereof for medicine.

## 31. Mountains in the sea.

Jon 2:6 I went down to the bottoms of the mountains; the earth with her bars was about me for ever: yet hast thou brought up my life from corruption, O LORD my God.

## 32. Expanding Universe.

Zec 12:1 The burden of the word of the LORD for Israel, saith the LORD, which stretcheth forth the heavens, and layeth the foundation of the earth, and formeth the spirit of man within him.

## 33. Origin of sexes explained.

Mar 10:6 But from the beginning of the creation God made them male and female.

Mar 10:7 For this cause shall a man leave his father and mother, and cleave to his wife;

Mar 10:8 And they twain shall be one flesh: so then they are no more twain, but one flesh.

## 34. Hint at Spherical Earth.

Luk 17:34 I tell you, in that night there shall be two men in one bed; the one shall be taken, and the other shall be left.

Luk 17:35 Two women shall be grinding together; the one shall be taken, and the other left.

Luk 17:36 Two men shall be in the field; the one shall be taken, and the other left.

## 35.Promiscuity Harmful.

1Co 6:18 Flee fornication. Every sin that a man doeth is without the body; but he that committeth fornication sinneth against his own body.

## 36. Ideal time for Circumcision.

Php 3:5 Circumcised the eighth day, of the stock of Israel, of the tribe of Benjamin, an Hebrew of the Hebrews; as touching the law, a Pharisee;

## 37. Medicinal use for Alcohol.

1Ti_5:23 Drink no longer water, but use a little wine for thy stomach's sake and thine often infirmities.

**38.2nd Law of Thermodynamics.**

Heb 1:11 They shall perish; but thou remainest; and they all shall wax old as doth a garment;

**39. Light Bursts created at conception of a person.**

Joh 1:9 *That* was the true Light, which lighteth every man that cometh into the world.

https://eternal-productions.org/101science.html

# Why Couldn't God Have Used Evolution

**275. The Bible Verses Evolutionary theory**

7 Differences (but not all of them) between Genesis 1 creation account and Evolution.

1. **Bible** has: Earth first. **Evolution** has: Sun first.

2. **Bible** has: light first. **Evolution** has: Sun first.

3. **Bible** has: Oceans before land. **Evolution** has: Land then Ocean.

4. **Bible** has: Plants then Marine creatures. **Evolution** has: Fish then plants.

5. **Bible** has: Plants then Sun. **Evolution** has: Sun then Plants.

6. **Bible** has: Marine Mammals then Land Mammals. **Evolution** has: the opposite.

7. **Bible** has: Birds then reptiles.

**Evolution** has: reptiles then birds.

Kent Hovind Creation Seminar 1 1hr 18min

# Other passages of scripture that conflicts with evolution

_Animals bring for after their kind._

Gen 1:24 And God said, Let the earth bring forth the living creature after his kind, cattle, and creeping thing, and beast of the earth after his kind: and it was so.

_No Predation in the beginning._

Gen 1:30 And to every beast of the earth, and to every fowl of the air, and to every thing that creepeth upon the earth, wherein _there is_ life, _I have given_ every green herb for meat: and it was so.

Gen 1:31 And God saw every thing that he had made, and, behold, _it was_ very good. And the evening and the morning were the sixth day.

_Genesis 5 lifespans._

_God made the world in a literal six days as an example to follow._

Exo 20:11 For _in_ six days the LORD made heaven and earth, the sea, and all that in them _is,_ and rested the seventh day: wherefore the LORD blessed the sabbath day, and hallowed it.

_People were larger._

Num_13:33 And there we saw the giants, the sons of Anak, _which come_ of the giants: and we were in our own sight as grasshoppers, and so we were in their sight.

_People over 2000 years ago live the same length as today._

Psa_90:10 The days of our years _are_ threescore years and ten; and if by reason of strength _they be_ fourscore years, yet _is_ their strength labour and sorrow; for it is soon cut off, and we fly away.

_No new information is being added to the gene pool. The genes just get shuffled._

Ecc 1:9 The thing that hath been, it *is that* which shall be; and that which is done *is* that which shall be done: and *there is* no new *thing* under the sun.

Ecc 1:10 Is there *any* thing whereof it may be said, See, this *is* new? it hath been already of old time, which was before us.

*Evolutionist also think we came from a rock.*

Jer 2:27 Saying to a stock, Thou *art* my father; and to a stone, Thou hast brought me forth: for they have turned *their* back unto me, and not *their* face: but in the time of their trouble they will say, Arise, and save us.

*Seems to imply we cannot just will changes to our bodies.*

Jer_13:23 Can the Ethiopian change his skin, or the leopard his spots? *then* may ye also do good, that are accustomed to do evil.

*When referencing Adam and Even Jesus said they were made at the beginning, not billions of years later.*

Mat_19:4 And he answered and said unto them, Have ye not read, that he which made *them* at the beginning made them male and female,

Mar_10:6 But from the beginning of the creation God made them male and female.

*Luke 3 Genealogy.*

*Jesus referenced Genesis, do not doubt Jesus.*

Joh_3:12 If I have told you earthly things, and ye believe not, how shall ye believe, if I tell you *of* heavenly things?

Joh 5:46 For had ye believed Moses, ye would have believed me: for he wrote of me.

Joh 5:47 But if ye believe not his writings, how shall ye believe my words?

*Man brought death into the world.*

Rom 5:12 Wherefore, as by one man sin entered into the world, and death by sin; and so death passed upon all men, for that all have sinned:

1Co 15:21 For since by man *came* death, by man *came* also the resurrection of the dead.

1Co 15:22 For as in Adam all die, even so in Christ shall all be made alive.

*Evolution is a fable that cannot tell us who the first humans were.*

1Ti 1:4 Neither give heed to fables and endless genealogies, which minister questions, rather than godly edifying which is in faith: *so do.*

*Evolution is falsely called science, but in reality religion.*

1Ti 6:20 O Timothy, keep that which is committed to thy trust, avoiding profane *and* vain babblings, and oppositions of science falsely so called:

1Ti 6:21 Which some professing have erred concerning the faith. Grace *be* with thee. Amen.

*2nd law thermodynamics defies upwards and on-wards evolution.*

Heb 1:11 They shall perish; but thou remainest; and they all shall wax old as doth a garment;

Heb 1:12 And as a vesture shalt thou fold them up, and they shall be changed: but thou art the same, and thy years shall not fail.

*Uniformitarianism predicted, Global flood deniers predicted.*

2Pe 3:2 That ye may be mindful of the words which were spoken before by the holy prophets, and of the commandment of us the apostles of the Lord and Saviour:

2Pe 3:3 Knowing this first, that there shall come in the last days scoffers, walking after their own lusts,

2Pe 3:4 And saying, Where is the promise of his coming? for since the fathers fell asleep, all things continue as *they were* from the beginning of the creation.

2Pe 3:5 For this they willingly are ignorant of, that by the word of God the heavens were of old, and the earth standing out of the water and in the water:

2Pe 3:6 Whereby the world that then was, being overflowed with water, perished:

2Pe 3:7 But the heavens and the earth, which are now, by the same word are kept in store, reserved unto fire against the day of judgment and perdition of ungodly men.

## *Atheism*

Gen 1:1 In the beginning God created the heaven and the earth.

Job 12:7 But ask now the beasts, and they shall teach thee; and the fowls of the air, and they shall tell thee:

Job 12:8 Or speak to the earth, and it shall teach thee: and the fishes of the sea shall declare unto thee.

Job 12:9 Who knoweth not in all these that the hand of the LORD hath wrought this?

Job 12:10 In whose hand *is* the soul of every living thing, and the breath of all mankind.

**Psa 14:1 To the chief Musician,** *A Psalm* **of David.** The fool hath said in his heart, *There is* no God. They are corrupt, they have done abominable works, *there is* none that doeth good.

**Psa 53:1 To the chief Musician upon Mahalath, Maschil,** *A Psalm* **of David.** The fool hath said in his heart, *There is* no God. Corrupt are they, and have done abominable iniquity: *there is* none that doeth good.

Psa 100:3 Know ye that the LORD he *is* God: *it is* he *that* hath made us, and not we ourselves; *we are* his people, and the sheep of his pasture.

Joh 1:3 All things were made by him; and without him was not any thing made that was made.

Rom 1:18 For the wrath of God is revealed from heaven against all ungodliness and unrighteousness of men, who hold the truth in unrighteousness;

Rom 1:19 Because that which may be known of God is manifest in them; for God hath shewed *it* unto them.

Rom 1:20 For the invisible things of him from the creation of the world are clearly seen, being understood by the things that are made, *even* his eternal power and Godhead; so that they are without excuse:

Rom 1:21 Because that, when they knew God, they glorified *him* not as God, neither were thankful; but became vain in their imaginations, and their foolish heart was darkened.

Rom 1:22 Professing themselves to be wise, they became fools,

Rom 1:23 And changed the glory of the uncorruptible God into an image made like to corruptible man, and to birds, and fourfooted beasts, and creeping things.

# The Anthropic Principle

## 276. The Anthropic Principle Example 1: The Sun's Mass

Too great: The suns brightness would change to fast and burn to quickly.

Too Small: Not enough ultraviolet radiation for plants to make sugars as well as oxygen.

The Evolution Handbook by Vance Ferrell pg 958

## 277. The Anthropic Principle Example 2: The Sun's Color

Too Red: photosynthetic response not enough.

Too Blue: phytosynthetic response not enough.

The Evolution Handbook by Vance Ferrell pg 958

## 278. The Anthropic Principle Example 3: Distance from the Sun

Too Close: planet too hot for a stable water cycle.

Too Far: our planet would be to cold for a stable water cycle.

The Evolution Handbook by Vance Ferrell pg 958

## 279. The Anthropic Principle Example 4: Earth's Gravity

Too Much: our water atmosphere would be to low.

Too Little: Atmosphere loses exceeding much water.

The Evolution Handbook by Vance Ferrell pg 958

## 280. The Anthropic Principle Example 5: Earth's Rotation

Too Fast: Too Vast of difference in diurnal temperature

Too Slow: wind too fast.

The Evolution Handbook by Vance Ferrell pg 958

## 281. The Anthropic Principle Example 6: Weak Nuclear Force

Too Little: Not enough hydrogen.

Too Much: Would convert to much hydrogen into helium.

The Evolution Handbook by Vance Ferrell pg 957

## 282. The Anthropic Principle Example 7: Strong Nuclear Force

Too Large: No hydrogen.

Too Small: Nothing but hydrogen.

The Evolution Handbook by Vance Ferrell pg 957

## 283. The Anthropic Principle Example 8: Electron to Proton Mass Ratio.

If just slightly larger or small this would cause inadequate chemical bonding.

The Evolution Handbook by Vance Ferrell pg 957

## 284. The Anthropic Principle Example 9: Oxygen

If there was too much oxygen, we would have a lot of trouble with fire.

Not enough oxygen and necessary chemical reactions would have a harder time happening.

Another important aspect of oxygen is it's dissolvability and it's speed of diffusing in water. These two things are crucial for our survival. If the dissolvability of oxygen was less, it couldn't be taken from an aqueous solution fast enough for metabolic requirements. Also note that breathing would be much harder if not for such a low viscosity and density of air. Let's also remember how plants give us oxygen and we give them carbon.

The Evolution Handbook by Vance Ferrell pg 952-954

## 285. The Anthropic Principle Example 10: Right Planetary Mass.

This allows the Earth to have the correct type and thickness of it's atmosphere. If the Earth was tinier, solar wind would tear up our atmosphere.

https://intelligentdesign.org/articles/list-of-fine-tuning-parameters/

*Warning: site may have an old universe view!*

## 286. The Anthropic Principle Example 11: Water

Unlike virtually all other substances that contract when they get cooler, water actually starts to expand when temperature gets below 39.2 F°. Then the Water rises to the surface then turns to Ice, the ice is lighter than the water below so it floats. Otherwise if it sank, more and more water would freeze until our oceans would all be frozen. There are several other properties of water which can be used as examples of the anthropic principle that I will not go into great detail here, I will let the reader do further research elsewhere. The reference that I provide below also contains more details on water including it's specific heat, Latent heat, surface tension, solvency, Thermal conductivity, reactivity, viscosity, and more.

The Evolution Handbook by Vance Ferrell pg 947-51

## 287. The Anthropic Principle Example 12: Light

Gen 1:3 And God said, Let there be light: and there was light.

Gen 1:4 And God saw the light, that *it was* good: and God divided the light from the darkness.

Gen 1:5 And God called the light Day, and the darkness he called Night. And the evening and the morning were the first day.

Light is among the electromagnetic spectrum. The range of total electromagnetic wavelengths is an astounding $10^{25}$. That's a big number. The majority of these wavelengths are very dangerous to life, thankfully (to Jesus), there are a number of barriers that block out these dangerous wavelengths. Among these is the ozone layer, the magnetic belts, and the atmospheric water vapor. These barriers protect us from X-rays, gamma rays, and microwaves; as well as protect us from the hazardous kinds of ultraviolet and infrared radiation. Another remarkable "coincidence" is the size of the aperture (entrance hole of the eye), the distance from the aperture to the retina (located at the back of the eye), and the wavelength of the radiation are just right to see clearly. If light's wavelength was just 5 microns bigger our eyes would have to be bigger than our entire face. Another great thing is that water is transparent to light, imagine how hard it would be for aquatic life if it wasn't. Could life in the ocean be possible without water being transparent to light?

The Evolution Handbook by Vance Ferrell pg 945-46

## 288.The Anthropic Principle Example 13: Polarity of Molecule (of Water)

Too great: vaporization and warmth of fusion too much for life to be sustainable.

Too little: Vaporization and warmth of fusion too little for life. Water wouldn't be solvent enough, and everything would freeze.

The Evolution Handbook by Vance Ferrell pg 958

# Final Remarkable Finds and Last Points

**289. Carl Baugh's Hyber-baric pressure research supports Genesis account.**

Geophysicist believe that the earth has expanded in diameter.

Gen 10:25 And unto Eber were born two sons: the name of one *was* Peleg; for in his days was <u>the earth divided</u>; and his brother's name *was* Joktan.

Prior to the days of Peleg and the flood of Noah, it has been calculated that the earth was 10 to 12% smaller in diameter. This would theoretically double our atmospheric pressure, by giving us a greater gravitational pull at sea level. By personal experience Dr. Baugh claims that it is much easier to breath under those conditions (greater atmospheric pressure), and by Doubling the atmospheric pressure this triples the assimilation of available oxygen. Dr. Baugh gives a very interesting testimony about his wife who had an open wound (which should have taken at least two weeks), was healed in three quarters of an hour. This explains the long life spans of pre-flood peoples.

Under two atmospheres of pressure (along with other conditions) Dr. Carl Baugh has...

- Altered the structure of the venom of snakes, which has theoretically nullified it.

· Tripled the adult lifespan of fruit flies, which is similar to humans living to be 200 years old, this even granting our degraded DNA.

(See point #43 and #180)

Hyperbaric Research Pressure That Heals Dr. Carl Baugh 33minutes to about 48 minutes

## 290. Earth's Magnetic field would be long gone.

The earth is losing it's magnetic field.

In the last hundred and fifty years 10% has been diminished, and 40% in the last thousands years!

Maximum Age: 25 Thousands years.

" Kent Hovind Creation Seminar 1, 1hr 21min"

## 291. Evolutionism: The *religion* of despair, hate, and selfishness.

Evolution is a faith based system. Evolutionists believe the universe popped into existence of of nothing without God. They believe all the animals we see here today came from some complex chemical soup billions of years ago. Evolutionists believe dinosaurs went extinct over 60 million years ago. In this sense evolution is very much a religion.

So where does the despair and selfishness come in? Well if there is no God, then life has no purpose, human life has no real value, and there are eternal consequences for our actions. Atheism and evolution lays the foundation for all kinds of evil in this world, here are some examples.

**1. Racism/Nazi-ism**- According to evolution some men are more evolved than others, Hitler was responsible for killing millions of Jews because he thought they were less human.

**2. Idolatry**- If someone doesn't believe in God then they will naturally be inclined to do their own will, and not God's. They will not pray, not read their bible, and not serve the true God. Instead they will serve football, money, or some other idol.

**3. Eugenics**- A highly popular motto of evolution is "survival of the fittest" and if man is just an advanced animal then why not kill handicapped people? Truly _evil_.

**4. Communism**-Many people tend to view mankind as a whole as basically good, this is one reason why communism doesn't work. The bible says (and therefore Christians affirm) that mankind is evil, it takes the inner working of the Holy Spirit to truly change a person from the inside out. Communism assumes that everyone would work for the benefit of society, this ideal could not be further from the truth, the lazy will take advantage of the diligent resulting in the impoverishment of the hard working and the promotion of sloth.

**5. Shoplifting**- I am willing to assume that a lot people believe that shoplifting is a victimless crime. If God's not real and this life is all we got and you could get away with it, why not shoplift? Without God we are just advanced clumps of cells with not purpose in a purposeless universe, there is no good and no evil.

**6. Pornography**- If Christianity is false then marriage is most likely just a social convention, it is the bible that defines what marriage is and if the bible is false, then why get married? Marriage takes a lot of work and is a lot of responsibility. Pornography is a lot easier and a lot cheaper than getting married and raising a family.

**7. Abortion**- The reason why humans have such immense value is because they are made in the image of the priceless, God himself. If humans are just descendants of apes or ape like creatures then why not abort them for convenience.

Can an atheists behave good (relative to a lot of humanity) probably. That is he surely can give to charity, work hard, help people with their lawn. However he has no foundation to do so according to his worldview. Love is not a material thing, therefore under naturalism, love doesn't exists.

Can a Christian do bad things, the possibility is surely there. Jesus was the only sinless one. When a christian sins he is behaving contrary to his worldview and is being inconsistent.

Suggested Video Hovind Seminar 5

David Wood Testimony

## 292. Lochness/plesiosaur Sightings, Zuiyo-maru carcass

Over the years there have been numerous reports of the Lochness monster sightings in a Lake in Scotland. The reports describe a creature that resembles what we would call a plesiosaur. When the road was put in beside the lake Lochness there were 52 sightings of something big and unidentifiable in the year 1933 alone. There have been a total of **3000** recorded sightings of Nessie, out of a total of **9000** sightings (around the time of the 1960's). I realize that there are probably some fake accounts of Nessie and other plesiosaur sightings out there but just because there are some false reports it doesn't rule out the real ones. If just 1 of the 9000 + reported sightings are legit, that would annihilate the theory of evolution. If i was a betting man I'd say some of those people saw a dinosaur. The video listed below covers more than 10 minutes of reported sightings and relevant subject matter. There is even an account listed in the video/seminar of a _plesiosaur killing four teenagers_ back in March of 1962 down in Pensacola, Florida.

The Zuiyo-maru carcass was a creature hauled up unto a Japanese fishing boat way back in 1977, it smelled terrible, weighted an astounding 2 Tonnes and was over 30 feet long, some believe it was a plesiousaur, others say it was a basking shark.

The Evolution Handbook by Vance Ferrell pg 481, "Hovind Seminar 3" 1hr 30min

## 293. Evolutionary art.

Isn't it interesting that when trying to prove Evolution, many times scientist or teachers will you illustrations, drawings, sculptures, in other words evidence that may not actually come from the real world but instead from the minds of man. Scientist and artist can only guess what the hair/skin color, as well as the ears, eyes, and lips were, the bones leave no clue. Contrast this with actual ancient artifacts we learned about earlier that mention biblical characters and locations, or observable science such as the kind law, Boyle's gas law, bio-genesis, etc and we see that it takes more faith to believe in evolution than Christianity.

The Evolution Handbook by Vance Ferrell pg 598-99

## 294.The fact that most scientists believe in evolution just verifies scripture.

The bible is clear that Satan has deceived the whole world (Rev 12:9), and that few find the path to life (Mat 7:14). Man is fallible, God's word is not. In 1770 it was claimed the world was 70 thousand years old, then in 1905 the earth was claimed to be 2 billion years old. Then again another age was given 3.5 billion years old, now it's 4.6 billion.

" Kent Hovind Creation Seminar 3", 45min

## 295. How did Kangaroos get to Australia?

The reason I mention this point is because I have heard this argument before that if all the animals got off of Noah's ark then how did the Kangaroos get to the Island or Continent rather of Australia?

There are two possible explanations. 1 They could have been transported by ships, not surprising because even in ancient times people brought animals by ship (see 2Chro 9:21). Another possibility is that they could have migrated there by land bridges before the sea level rose to it's present state.

## 296. Sauropod Dino in Congo river sightings (Mokele-mbembe).

Recent Sightings of a sauropod dinosaur (Apatosaurus) has been spotted by the natives down in the swamp of Africa. A retired missionary names Eugene Thomas claimed that two pygmies of his church killed one way back in 1959.

" Kent Hovind Creation Seminar 3" 1r 23min

## 297. Modern Pterodactyle sightings.

Referred to as the Batamzinga Kongamato. These creature was reported to be still living by "Steve" Romandi ( a former LSU student in Baton Rouge, LA). He claimed that they lived in his village in Kenya.

Additionally, Dr Carl Baugh has interviewed the natives of Papua, New Guinea. They have reported also seeing Pterodactyles, which they call Ropen. They reportedly are said to glow in the dark.

" Kent Hovind Creation Seminar 3" 1hr 46min

**298. Public schools are teaching kids what to think instead of how.**

E.g. Giving children _loaded questions_ like, Do you think humans are still evolving? This is what is known as indoctrination, not education. It is also a logical fallacy.

" Kent Hovind Creation Seminar 4, 1hr 07min"

**299. Evolutionist try to use micro-evolution to prove the other five types of Evolution.**

The first five types of evolution are religious not science. In fact the first five types of evolutionary theory are impossible (many reasons why are giving in this book).

**1. Cosmic Evolution**- Origin of space, time, matter.

**2. Chemical Evolution**- origin of elements other than hydrogen.

**3. Stellar Evolution**- origin of Stars and planets.

**4. Organic Evolution**- Origin of Life

**5. Macro Evolution**- Changing into other animals.

" Kent Hovind Creation Seminar 4, 44min"

**300.The Evidence from Conversions.**

And last, but certainly not least we have the countless, numerous testimonies of those who's lives that Jesus has touched. I invite my reader to listen to the testimonies on youtube of former unbelievers who have had encounters with the risen saviour.

Here is a small sample of *testimonies* that I think would give skeptics and atheists pause to reconsider their position.

1.  *Will Vining (Former OCD Victim Shares Powerful Tesimony!)
2.  *Isaiah Saldivar (I Was an Atheist Until This Happened...(Testimony))
3.  John Ramirez-Ex-Satanist
4.  Nicholas Bowling
5.  *Richard Lorenzo Jr. (From Demonic Warlock in Training to Following JESUS!)
6.  David Wood- Former Atheist
7.  The LORD's Testimony in Tragedy: God Spared this Life//December 5, 2019 (Jesus is Life Ministries)
8.  Youtube Video: Story how God cured spine disease - Nick Vujicic

A couple Youtube channels that may prove helpful in searching for testimonies.

*Delafe Testimonies (contains dozens if not much more personal testimonies)

One for Israel (Evangelizes Jews and Arabs for Jesus, contains testimonies)


**The End.**

# Don't miss out!

Visit the website below and you can sign up to receive emails whenever Justin Horn publishes a new book. There's no charge and no obligation.

https://books2read.com/r/B-A-KAVIB-YTTID

**BOOKS2READ**

Connecting independent readers to independent writers.

www.ingramcontent.com/pod-product-compliance
Lightning Source LLC
Chambersburg PA
CBHW061531120726
48001CB00004B/1488